Conceptual Approach to Quantum Electrodynamics

Online at: https://doi.org/10.1088/978-0-7503-6054-8

Conceptual Approach to Quantum Electrodynamics

Samina S Masood
Department of Physics, University of Houston Clear Lake, Houston, TX, USA

IOP Publishing, Bristol, UK

ISBN 978-0-7503-6054-8 (ebook)
ISBN 978-0-7503-6052-4 (print)
ISBN 978-0-7503-6055-5 (myPrint)
ISBN 978-0-7503-6053-1 (mobi)

DOI 10.1088/978-0-7503-6054-8

Version: 20251101

IOP ebooks

British Library Cataloguing-in-Publication Data: A catalogue record for this book is available from the British Library.

Published by IOP Publishing, wholly owned by The Institute of Physics, London

IOP Publishing, No.2 The Distillery, Glassfields, Avon Street, Bristol, BS2 0GR, UK

US Office: IOP Publishing, Inc., 190 North Independence Mall West, Suite 601, Philadelphia, PA 19106, USA

I dedicate this work to the Almighty, first to let me execute a prime dream. In addition, I dedicate this work to my parents and my spouse whose continuous encouragement and support made this work possible.

Contents

Preface

Since I was a student, I always wished for a single, comprehensive book to revisit previously learned subjects that tend to fade over time. Motivated by this desire, I aimed to write a book that consolidates essential physical concepts and fundamental laws of physics, enriched by deeper conceptual explanations than typically found in undergraduate textbooks.

The primary purpose of this book is to provide an accessible introduction for incoming graduate students, guiding them through the evolution of quantum field theory, in particular, quantum electrodynamics (QED). It emphasizes the conceptual growth of the theory within the broader human quest to unravel the mysteries of the universe. To fulfill this long-standing ambition, I have reviewed crucial topics comprehensively, presenting fundamental equations alongside their conceptual significance in the context of QED. This text serves as a practical resource, helping beginners refresh their existing knowledge and bridge any gaps in their understanding. Essential mathematical relationships are revisited to establish a solid technical foundation, complemented by reviews of electromagnetism, quantum mechanics, relativity, and high-energy physics, all crucial for grasping quantum field theory.

Following the exploration of quantum field theory's development and a brief introduction to QED, this book further highlights the significance of QED in applied physics, illustrating its importance in technological advancement and innovation.

Acknowledgements

The author is extremely grateful to the Creator for providing the opportunity and inspiration to fulfill this long-desired project in collaboration with IOP Publishing. Heartfelt gratitude is extended to my family, particularly to my spouse, whose unwavering support and patience have been invaluable throughout this journey. I am especially thankful to my former student and current colleague, Mr. Derek Smith, for meticulously reviewing and proofreading the entire manuscript and for providing constructive feedback. I also express sincere appreciation to Dr. S Caliskan and Dr. Hussain Rizvi for their insightful recommendations. The encouragement and support from my workplace were indispensable and deeply appreciated. Lastly, special thanks go to the dedicated team at IOP Publishing, whose professionalism, guidance, and patience significantly contributed to the successful completion of this project.

Author biography

Samina Masood

The author **Dr S Masood** is a professor of Physics at the University of Houston Clear Lake (UHCL). She got her PhD from Quaid-i-Azam University, Islamabad, Pakistan in high-energy physics and astrophysics. Her PhD work was based on quantum statistical field theory and its applications to the thermal history of the Universe and the interior of stars.

In 1998, she relocated to the USA due to family obligations, resulting in an interruption in her research activities which were additionally exacerbated by personal challenges, including health issues. In 2006, she resumed her academic career by joining the University of Houston-Clear Lake. After an initial lull in research, her interest in biophysics gradually intensified, catalyzed by the intriguing applications of perturbation theory in quantum mechanics to biomolecules. This part-time research eventually expanded her appreciation of the broader applications of quantum mechanics in molecular biology.

Despite heavy teaching responsibilities and other constraints, she continued research in astroparticle physics, driven by her passion for teaching and her enduring interest in scientific inquiry. Her academic commitment culminated in the creation of this book, which synthesizes her teaching and research experiences into a comprehensive educational resource.

Part I

Mathematical background

IOP Publishing

Conceptual Approach to Quantum Electrodynamics

Samina S Masood

Chapter 1

Vectors and tensors

1.1 Introduction

Vector quantities are defined by their magnitudes and directions. A complete set of orthogonal unit vectors or basis vectors generate a vector space. Each vector in a vector space is expressed as a linear combination of all the components represented by the orthogonal basis vectors. Vector spaces are generated by a complete set of mutually independent basis vectors and the components of vectors along the basis vectors are called coordinates. Each component of a vector corresponds to a basis vector of the corresponding vector space, whereas the same vector in another vector space (coordinate system) is described with a different set of components in terms of the corresponding basis vectors without changing the magnitude of the original vector. Therefore, vectors can be transformed from one coordinate system to another coordinate system by transformation of each and every component without changing their magnitudes. The magnitude of a vector can always be calculated from the square root of the sum of squares of each component in a particular vector space.

These vectors can be transformed to other vectors when operated or multiplied with an appropriate transformation square matrix. Multiplication with a square matrix transforms one vector into another vector (e.g., rotation) or transforms from one to another vector space (e.g., Cartesian to spherical polar coordinates). Properties of the transformation matrices determine the properties of the vector space as well. A set of all the transformation matrices of a vector space can make a group if it makes a closed set including an identity matrix and if every matrix has an inverse. A group of unitary matrices corresponds to a fundamental interaction as it satisfies the requirements of conservative forces as well.

doi:10.1088/978-0-7503-6054-8ch1

1.2 Vector calculus

A vector is usually represented in bold face letter **V** (or $\overrightarrow{V}$) in a three-dimensional space, generated by three orthogonal basis vectors ($\hat{e}_1$; $\hat{e}_2$; and, $\hat{e}_3$), which are also called unit vectors. The corresponding components of any vector along these basis vectors are identified as V_1, V_2 and V_3. For example, the well-known unit vectors of Cartesian coordinates ($\hat{i}, \hat{j}, \hat{k}$) and the spherical polar coordinates, ($\hat{r}$, $\hat{\theta}$, $\hat{\phi}$) are the basis vectors generating the corresponding three-dimensional coordinate spaces, respectively. The coordinates, associated with these basis vectors are sometimes identified as generalized coordinates. A unit vector along the direction of a vector **V** is represented as:

$$\hat{V} = \frac{\mathbf{V}}{|\mathbf{V}|} \tag{1.1}$$

A vector can then be expressed as a linear combination of all its components along the direction of its basis vectors as:

$$\begin{aligned}\mathbf{V} &= (\mathbf{V}\cdot\mathbf{e}_1)\hat{\mathbf{e}}_1 + (\mathbf{V}\cdot\mathbf{e}_2)\hat{\mathbf{e}}_2 + (\mathbf{V}\cdot\mathbf{e}_3)\hat{\mathbf{e}}_3\\ &= V_1\hat{\mathbf{e}}_1 + V_2\hat{\mathbf{e}}_2 + V_3\hat{\mathbf{e}}_3\end{aligned}$$

The dot product of a vector with its basis vectors $\hat{\mathbf{e}}_\mathbf{n}$ in an n-dimensional space, gives the projection of the vector along the nth basis vector and may be called the nth component of the vector along the unit vector $\hat{e}_n$. These components of a vector V_n lie along the basis vectors $\hat{e}_n$ and determine the direction of orientation of the vector in space. The direction cosines of a vector **V** are the ratios of the components of a vector with the magnitude of the vector. The direction cosine of the nth basis vector $\hat{e}_n$ for each n is defined as $\frac{\mathbf{V}\cdot\mathbf{e_n}}{|V|}$. The dot product of two vectors can be seen as a projection of one vector along the other or vice versa, whereas the cross product gives a vector (as a normal vector) coming out of the plane of two vectors (generated by the Kronecker product of both vectors) in the perpendicular direction.

1.2.1 Vector identities

Some of the common vector identities in three-dimensional vector spaces are given as:

$$\begin{aligned}&\mathbf{A}\cdot\mathbf{B} = \mathbf{B}\cdot\mathbf{A}\\ &\mathbf{A}\times\mathbf{B} = -\mathbf{B}\times\mathbf{A}\\ &\mathbf{A}\cdot(\mathbf{B}+\mathbf{C}) = \mathbf{A}\cdot\mathbf{B}+\mathbf{A}\cdot\mathbf{C}\\ &\mathbf{A}\cdot\mathbf{B} = A_1B_1 + A_2B_2 + A_3B_3\\ &\mathbf{A}\times(\mathbf{B}+\mathbf{C}) = \mathbf{A}\times\mathbf{B}+\mathbf{A}\times\mathbf{C}\\ &\mathbf{A}\times(\mathbf{B}\times\mathbf{C}) = \mathbf{B}(\mathbf{A}\cdot\mathbf{C}) - \mathbf{C}(\mathbf{A}\cdot\mathbf{B})\\ &\mathbf{A}\cdot(\mathbf{B}\times\mathbf{C}) = \mathbf{B}\cdot(\mathbf{C}\times\mathbf{A}) = \mathbf{C}\cdot(\mathbf{A}\times\mathbf{B})\\ &(\mathbf{A}\times\mathbf{B})\cdot(\mathbf{C}\times\mathbf{D}) = (\mathbf{A}\cdot\mathbf{C})(\mathbf{B}\cdot\mathbf{D}) - (\mathbf{A}\cdot\mathbf{B})(\mathbf{B}\cdot\mathbf{C})\\ &\mathbf{A}\times\mathbf{B} = (A_2B_3 - A_3B_2) + (A_3B_1 - A_1B_3) + (A_1B_2 - A_2B_1),\end{aligned} \tag{1.2}$$

where **i, j, k** are the unit vectors in x-, y- and z-directions, respectively. A product of two vectors in the same vector space can produce a symmetric and an antisymmetric combination of two vectors. A symmetric combination or a dot product is a projection of any one of these vectors on the other vector. It is a scalar quantity and can be considered as a scalar relation of two vectors. The antisymmetric combination or the product of two vectors gives a vector perpendicular to the plane generated by both vectors and the direction of the vector depends on the order of multiplication of vectors and is named as a vector product. Therefore, the product of two vectors is expressed in two terms that give a sum of a dot product and a cross product. The dot product gives the sum of the product of parallel components calculated from a sum of diagonal elements, called the trace of a diagonalized matrix, and the cross product gives the antisymmetric combination of perpendicular components in the given space.

A cross product of three-dimensional vectors (in three-dimensional space) is written as a product of three vectors $\hat{\mathbf{A}} \times (\hat{\mathbf{B}} \times \hat{\mathbf{C}})$ in the direction of a unit vector $\hat{e}$ in the Cartesian coordinates. Three-dimensional volume of a parallelepiped (in three-dimensional space) is expressed as a determinant of the resultant square matrix in three dimensions $\hat{\mathbf{A}} \cdot (\hat{\mathbf{B}} \times \hat{\mathbf{C}})$. The diagonal elements of this matrix give various terms of a scalar product. Off-diagonal elements of this matrix represent components of a cross product, and a matrix M shows all the component in higher dimensions.

$$\begin{aligned} \mathbf{AB} &= \mathbf{A}\cdot\mathbf{B} + \mathbf{A}\times\mathbf{B} + M \\ \text{and}\quad \mathbf{BA} &= \mathbf{A}\cdot\mathbf{B} + \mathbf{B}\times\mathbf{A} + M^T \end{aligned} \tag{1.3}$$

while,

$$(\mathbf{B}\times\mathbf{A} = -\mathbf{A}\times\mathbf{B})$$

M is a square matrix and M^T is its transpose.

In linear algebra, the diagonalization procedure is used to find the basis vectors. Vanishing of off-diagonal components insures the orthogonality of basis vectors. Antisymmetric vector products are associated with the product of two vectors perpendicular to the plane generated by two multiplying vectors in a three-dimensional coordinate system. Angular momentum $\mathbf{L} = \mathbf{r} \times \mathbf{p}$ is a combined effect of moment arm $\vec{r}$ and three-dimensional momentum $\vec{p}$ perpendicular to a plane generated by a cross product of both vectors. Similarly, torque is generated by a product of $\vec{r}$ (i.e. the distance between the point of application of the force and the axis of rotation) and the applied force $\vec{F}$ perpendicular to $\vec{r}$. The antisymmetric nature of a vector product is related to the order of multiplication of two vectors. Scalar quantities are produced by the dot product of two vectors, and the vectors lose directionality in a dot product.

The variation of vectors in a vector space is described as a vector itself. The components of a vector are written as a sum of projections of the vector along the corresponding coordinates (basis vectors). These components contribute to the magnitude of a vector and cannot be changed independently because the total magnitude of vectors is related to all components and has a fixed ratio of each component as a direction cosine with the total magnitude. However, the derivatives and integrals of vectors measure the change in magnitude and may just occur along one basis vector without affecting the ratios of the other components. Orthogonality of basis vectors allows a vector to change along one basis vector while keeping every other component constant. This is mathematically given as a partial derivative (e.g., moving just along a straight line).

The vector operator $\vec{\nabla}$ then describes the variation of vectors along each basis vector in the form of partial derivatives and is expressed in the components form, calculating the change of vector along each direction and expressing it as a linear combination of all basis vectors. The vector operator $\vec{\nabla}$ can operate on a vector to describe its change in space in terms of the variation in every component in the corresponding basis vector and is treated as a vector in itself. In commonly used three-dimensional coordinate systems (Cartesian, polar and cylindrical), $\vec{\nabla}$ or simply ∇ is defined as:

$$\begin{aligned}
\nabla &= \frac{\partial}{\partial x}\hat{i} + \frac{\partial}{\partial x}\hat{j} + \frac{\partial}{\partial x}\hat{k} \quad \text{.........(a)} \\
\nabla &= \hat{e}_1\frac{\partial}{\partial r} + \hat{e}_2\frac{1}{r}\frac{\partial}{\partial \theta} + \hat{e}_3\frac{1}{r\sin\theta}\frac{\partial}{\partial \phi} \quad \text{.........(b)} \\
\nabla &= \hat{e}_1\frac{\partial}{\partial \rho} + \hat{e}_2\frac{1}{\rho}\frac{\partial}{\partial \phi} + \hat{e}_3\frac{\partial}{\partial z} \quad \text{.........(c)}
\end{aligned} \tag{1.4}$$

in the usual notation of these coordinate systems, respectively. These vectors can be expressed in any three-dimensional commonly known vector spaces such as the Cartesian coordinate, the spherical polar coordinates and the cylindrical coordinates in terms of the corresponding basis vectors of the coordinate systems. $\vec{\nabla}$ is a three-dimensional vector operator and is only defined in the three particular forms of equation (1.4). These forms are obtained using the total derivative of a function; using a general form of coordinates x_1, x_2, x_3 instead of x, y, z coordinates such that:

$$df(x, y, z) \equiv df(x_1, x_2, x_3) = \frac{\partial f}{\partial x_1}dx_1 + \frac{\partial f}{\partial x_2}dx_2 + \frac{\partial f}{\partial x_3}dx_3$$

This relation can be used to evaluate a total derivative with respect to any general variable z, such that

$$\frac{df(x, y, z)}{dz} = \frac{\partial f}{\partial x_1}\frac{dx_1}{dz} + \frac{\partial f}{\partial x_2}\frac{dx_2}{dz} + \frac{\partial f}{\partial x_3}\frac{dx_3}{dz}$$

Differentiation or integration of vectors could either be a scalar or a vector operation depending on its operation along a particular component or in the entire vector space, evaluated as the orientation along each of the basis vectors, respectively. It is scalar if it treats components as variables, related to the magnitude of the components and ignoring the variation of other components. The scalar form of differential operations extracts the contribution of one component at a time, keeping the rest unchanged. It may therefore isolate the variation in one coordinate of interest, getting rid of the contribution of all other variables due to orthogonality.

The three-dimensional differential operator is a vector operator and is written in terms of its basis vectors and its components are partial derivatives of vectors with respect to one component at a time, keeping every other component constant, This three-dimensional vector operator of vectors is represented as the 'del' operator, which is a mathematical operation describing the variation of a vector in the vector space. All the components of vectors and the components of differential operators are in the same vector space. Due to the vector nature of this operator, it affects the vector properties. Therefore, the vector operation $\vec{\nabla}$ on a vector changes the magnitude as well as the orientation of the vector in the same vector space. The vector product of two vectors gives an array of elements due to the product of one vector on another and changes its direction if the operated vector is changed.

The application of a vector operator on a scalar imposes vector behavior on a scalar function of spacial coordinates. This operation called a gradient is a vector operator that changes a scalar into a vector function, e.g., $(\vec{\nabla}\phi)$ makes a scalar field $\phi(x)$ vary differently in different directions and the variation of function in space induces a vector characteristic. A few well-known examples are gradients of temperature, pressure and density as these scalars change in space depending on their direction. These gradients are very useful mathematical operators to determine the dynamics of many-body systems such as fluids.

Differential operation on vectors can be performed in more than one way. A simple vector operation given as a dot product of ∇ with a vector is called the divergence of a vector and is similar to the dot product of two vectors. However, this product is written in a particular order and the differential operator should be operated from the left-hand side and is given in the components form as:

$$\nabla \cdot \mathbf{V} = \frac{\partial V_1}{\partial x_1} + \frac{\partial V_2}{\partial x_2} + \frac{\partial V_3}{\partial x_3} = \nabla_1 V_1 + \nabla_2 V_2 + \nabla_3 V_3$$

It basically takes away the vector nature and extracts the magnitude of variation of a vector in space as a gradient of the vector and represents it as a scalar which gives the net variation of all the components of the vector parallel to all the basis vectors. The components of a del operator ∇_1, ∇_2 and, ∇_3 are components of the differential operator along the three basis vectors. In a way, this operation is opposite to the gradient operation and extracts the net variation of the vector in space. On the other hand, the divergence operator is just like a dot product of ∇ with another vector. However, like other dot products, the divergence operator does not commute due to the non-commuting nature of the differential operator, $\nabla \cdot \mathbf{A} \neq \mathbf{A} \cdot \nabla$. Therefore, the order of differential operation is always important. Similarly, the vector product of ∇ with a vector is not antisymmetric due to the non-commuting nature of the components of the differential operator.

$$\nabla \times \mathbf{A} \neq -\mathbf{A} \times \nabla$$

The cross product of $\vec{\nabla}$ with a vector is called the curl of a vector and gives a resultant vector. The curl operation rolls a vector in three-dimensional space. Every component is associated with a net effect of the perpendicular components of all other components of force along each basis vector as a force is needed to roll it over. The following differential forms of vectors correspond to the (a) gradient, (b) divergence and (c) curl of vectors. These vectors can always be represented in the form of three components of a vector in the corresponding coordinate system and can be expressed in a general form of the components of a vector in the relevant coordinates $A_i = (A_1, A_2, A_3)$ as:

$$\nabla\phi = \frac{\partial\phi}{\partial x_1}\hat{\mathbf{e}}_1 + \frac{\partial\phi}{\partial x_2}\hat{\mathbf{e}}_2 + \frac{\partial\phi}{\partial x_3}\hat{\mathbf{e}}_3. \text{ ......(a)}$$

$$\nabla \cdot \mathbf{A} = \nabla_1 A_1 + \nabla_2 A_2 + \nabla_3 A_3. \text{ ......(b)}$$

$$\nabla \times \mathbf{A} = (\nabla_2 A_3 - \nabla_3 A_2)\hat{\mathbf{e}}_1 + (\nabla_3 A_1 - \nabla_1 A_3)\hat{\mathbf{e}}_2 + (\nabla_1 A_2 - \nabla_2 A_1)\hat{\mathbf{e}}_3. \text{ ......(c)}$$

A few commonly-used vector identities with a vector differential operator are:

$$\begin{aligned}
&\nabla \times \nabla \phi = \mathbf{0}\\
&\nabla \cdot (\nabla \times \mathbf{A}) = \mathbf{0}\\
&\nabla \cdot (\phi\mathbf{A}) = \mathbf{A} \cdot (\nabla\phi) + \phi(\nabla \cdot \mathbf{A})\\
&\nabla \times (\nabla \times \mathbf{A}) = \nabla(\nabla \cdot \mathbf{A}) - \nabla^2 \mathbf{A}\\
&\nabla \times (\mathbf{A} + \mathbf{B}) = \nabla \times \mathbf{A} + \nabla \times \mathbf{B}\\
&\nabla \times (\phi\mathbf{A}) = \nabla\phi \times \mathbf{A} + \phi(\nabla \times \mathbf{A})\\
&\nabla(\mathbf{A} \cdot \mathbf{B}) = (\mathbf{A} \cdot \nabla)\mathbf{B} + (\mathbf{B} \cdot \nabla)\mathbf{A}\\
&\nabla \cdot (\mathbf{A} \times \mathbf{B}) = \mathbf{B} \cdot (\nabla \times \mathbf{A}) - \mathbf{A} \cdot (\nabla \times \mathbf{B})\\
&\nabla \times (\mathbf{A} \times \mathbf{B}) = \mathbf{A}(\nabla \cdot \mathbf{B}) - \mathbf{B}(\nabla \cdot \mathbf{A}) + (\mathbf{B} \cdot \nabla)\mathbf{A} - (\mathbf{A} \cdot \nabla)\mathbf{B}
\end{aligned} \tag{1.5}$$

Integration of vectors is much more complicated than the integration of scalar functions. All three-dimensional vector operations are associated with the ∇ operator and are expressed in component forms that are scalars in themselves and the directional effect is studied separately. The integration of vectors is only possible in the presence of another vector such that either the variable of integration is a magnitude of a vector and its direction is separated as a unit vector or we take a dot product of this unit vector with another integrating vector to integrate the components. Basically, differential and integration operations are meant for scalar functions of variables. The vector nature remains unchanged by the spatial variation. The vector coponents of these operators appear as unit vectors which describe how these operations will be seen in different directions.

The cross product of a vector emerges as parallelepiped, perpendicular to the plane generated by two vectors. Now to understand the curl, imagine a three-dimensional flexible object and stretch it with an independent vector force that has different components along each axis and give it a random shape. This is basically what a curl operation does. Since it indicates different components of a force, its volume integral can be related to a line integral because stretching by a three-dimensional perpendicular force modifies its shape for a closed volume.

1.2.2 Integral identities

Both of these operations do not change the orientation of a vector. Some of the helpful integral theorems of vector calculus can be listed as:

$$\begin{aligned}\oint[L(x_1\cdot x_2)\,dx_1 + M(x_1\cdot x_2)\,dx_2] &= \iint\left[\frac{\partial \mathbf{M}}{\partial \mathbf{x_1}} - \frac{\partial \mathbf{L}}{\partial \mathbf{x_2}}\right]\cdot \hat{\mathbf{n}}\mathbf{da}\\ \int(\nabla\phi)\mathbf{dV} &= \int \phi\mathbf{da} = \int \phi\cdot\hat{\mathbf{n}}\mathbf{da}\\ \oint \mathbf{A}\cdot\mathbf{da} &= \int(\nabla\cdot\mathbf{A})\mathbf{dV} = \int(\mathbf{A}\cdot\hat{\mathbf{n}})\mathbf{da}\\ \oint \mathbf{A}\cdot\mathbf{dl} &= \int(\nabla\times\mathbf{A})\cdot\mathbf{da} = \int(\nabla\times\mathbf{A})\cdot\hat{\mathbf{n}}\mathbf{da}\end{aligned} \tag{1.6}$$

The unit vector $\hat{\mathbf{n}}$ is perpendicular to the area element **da**. A positive unit $\hat{n}$ vector is coming out of the area, while a negative unit vector goes inward. $L(x, y)$ and $M(x, y)$ are two functions in an xy-plane. The closed integrals indicate the wrapping of the area by a counterclockwise loop. If it is wrapped clockwise, it will give a negative sign making area a vector quantity.

The above equations are describing the key principles of vector calculus for physical applications. Equation (1.6) is usually referred to as Green's theorem and is very useful in describing the fluid equations and relates the line and surface integrals in two-dimensional space. The second and third equations of equations (1.6) relate the volume integral with the surface integral, and the third equation is known as Gauss's divergence theorem. The last equation of (1.6) relates the surface integral to the volume integral as well and is known as Stokes' theorem. These integral operations are considered to be very helpful in vector calculus because they help to reduce the order of integration or let you convert the integral into known variables. The angular area just like the circular area element of a conical section at a given distance r is given for the variation in angle between θ and $\theta + d\theta$ and for small $d\phi$ as $d\Omega = \sin\theta d\theta d\phi$ and the volume of the cone is given as:

$$dV \equiv r^2 dr d\Omega$$

where

$$d\Omega = \sin\theta d\theta d\phi$$

such that

$$o \leqslant \phi \leqslant \pi/2,\ o \leqslant \theta \leqslant 2\pi$$

Vector quantities are described as an array in space. The generalization of vectors can be described as tensors. Scalars are zero-rank tensors, vectors have rank 1 and matrices have rank 2. Matrices have two-dimensional arrays and can be shown on paper. Three-dimensional arrays or studies of higher dimensional arrays cannot be written on paper and a mathematical way to define them is evolved using the index form borrowed from the component form of matrices. Tensors provide an easy way to conceptually develop higher dimensional arrays and develop complicated algebra. At this point a collective notation can be introduced to handle higher-dimensional algebra. For this purpose vectors can be expressed in matrix notation and then those matrices can be written in a compact form in tensor notation.

The next section is devoted to the discussion of tensors. The matrix representation of vectors makes it convenient to describe tensors in matrix representation. Tensors basically emerge from transformation properties of vectors. Scalars are tensors of rank 0 due to their invariance from one coordinate space to another coordinate space. Three-dimensional coordinate space is the same space generated by three well-known sets of generalized coordinates or basis vectors, described as Cartesian coordinates, spherical polar coordinates and cylindrical coordinates. A vector in three-dimensional space can essentially be expressed by any set of three basis vectors in the corresponding coordinate space.

Vectors need a transformation matrix to go from one coordinate system to another coordinate system and are identified as tensors of rank 1. Scalars need no transformation matrix and remain invariant as zero-rank tensors. Matrices need two transformation matrices to change rows and columns independently and are tensors of rank 2. A product of two vectors can have a rank 2 or less. A tensor that can be reduced to lower rank tensors is called a reducible tensor, whereas one with a fixed rank is classified as an irreducible tensor. A combination of two vectors needs two transformation matrices (for two rank 1 tensors). As mentioned earlier, in reference to scalar and vector products, the gradient operator increases the rank of a tensor by one and divergence reduces it by one as divergence can be considered as a dot product of two vectors, which converts two vectors into a scalar quantity. The curl operation gives the effect of variation of a vector along its normal vector.

1.2.3 Generalization of vectors into matrices

Vectors are represented by arrays in a vector space giving all the components of a vector as elements of the array. All vector operations can be described in matrix notation. If we consider an n-dimensional vector space generated by a set of n number of basis vectors, a vector in this vector space is represented as an array of n matrix elements. Transformation of a vector from one vector space to another vector space of the same dimensions occurs through multiplication with a transformation (square) matrix. Tensors can therefore be considered as the most general transformation of a vector with n number of arrays and each array needs a transformation matrix to transform from one vector space to another. A general expression of a vector with a finite number of required $(n \times n)$ transformation matrices is called a tensor. We can understand tensors considering the generalization of vector transformation.

A vector is expressed in terms of the generalized coordinates q_j (for $j = 1, 2, 3, \ldots$) in a vector space. For example, a small displacement vector $d\mathbf{x}$ in a three-dimensional space can be written in the component form as:

$$d\mathbf{s} \equiv ds_i = \sum_j h_{ij} dq_j \hat{\mathbf{e}}_j$$

$$dx_i = \left(\frac{\partial x_i}{\partial q_1}\right) dq_1 \hat{\mathbf{e}}_1 + \left(\frac{\partial x_i}{\partial q_2}\right) dq_2 \hat{\mathbf{e}}_2 + \left(\frac{\partial x_i}{\partial q_3}\right) dq_3 \hat{\mathbf{e}}_3 \tag{1.7}$$

$$h_{ij} = \frac{\partial x_i}{\partial q_j}$$

where, i, j and k are indices in three-dimensional space and have the integral values $i, j = 1$, 2 and 3. All the above relations can be generalized to n basis vectors, just giving the variation to these indices as $1, \ldots, n$. In the general form, we can construct two matrices, the identity matrix δ_{ij} for $i = j$ and antisymmetric matrix $g_{ij} = 0$ for $i \neq j$ such that $g_{ii} = h_i^2$ and the shortest distance between two points is given as

$$d(\mathbf{s}\cdot\mathbf{s}) = ds^2 = \sum_i \delta_{ij} h_i h_j dq_i dq_j \hat{\mathbf{e}}_i \hat{\mathbf{e}}_j$$

and can be written in the component form in equation (1.8) as:

$$\begin{aligned} ds^2 &= g_{11} dq_1^2 + g_{12} dq_1 dq_2 + g_{13} dq_1 dq_3 \\ &+ g_{21} dq_2 dq_1 + g_{22} dq_2^2 + g_{23} dq_2 dq_3 \\ &+ g_{31} dq_3 dq_1 + g_{32} dq_3 dq_2 + g_{33} dq_3^2 \end{aligned} \tag{1.8}$$

This same expression, in the matrix form, becomes

$$\begin{bmatrix} d\vec{x_1} \\ d\vec{x_2} \\ d\vec{x_3} \end{bmatrix} = \begin{bmatrix} g_{11} & g_{12} & g_{13} \\ g_{21} & g_{22} & g_{23} \\ g_{31} & g_{32} & g_{33} \end{bmatrix} \begin{bmatrix} d\vec{q_1} \\ d\vec{q_2} \\ d\vec{q_3} \end{bmatrix} \tag{1.9}$$

which attains the compact matrix notation as:

$$ds^2 = \sum_{ij} g_{ij} dq_i dq_j = \sum h_i^2 dq_i^2 = h_1^2 dq_1^2 + h_2^2 dq_2^2 + h_3^2 dq_3^2 \tag{1.10}$$

where, indices is and js determine the matrix elements of the transformation matrix g_{ij} and can be expressed as:

$$g_{ij} = \frac{\partial x}{\partial q_i}\frac{\partial x}{\partial q_j} + \frac{\partial y}{\partial q_i}\frac{\partial y}{\partial q_j} + \frac{\partial z}{\partial q_i}\frac{\partial z}{\partial q_j} = \sum_l \frac{\partial x_l}{\partial q_i}\frac{\partial x_l}{\partial q_j} = \sum_l h_{il} h_{jl}$$

If we consider an incoming state as a column vector and an outgoing state as a row vector, the transformation of a row vector will be a conjugate matrix of the required transformation for the column vector. A conjugate matrix here is defined as a transpose matrix with all the elements as complex conjugates of the corresponding elements($a_{ji} = a_{ij}^*$). So we can talk about inverse transformation for row and column vectors as the transpose of each other for square matrices. Every vector needs a transformation matrix like g_{ij} and is therefore a tensor of rank 1.

These transformations are not limited to the coordinate systems only. Within the same coordinate system, the change of reference points or rotation around a fixed point gives practically a different set of basis vectors. Therefore, a transformation matrix is required to see the transformation of components. Rotation matrices provide a simple example of tensors. They tell us how the rotation about an axis by certain angle affects the components of a vector such that

$$\begin{bmatrix} x' \\ y' \end{bmatrix} = \begin{bmatrix} \cos\theta & \sin\theta \\ -\sin\theta & \cos\theta \end{bmatrix} \begin{bmatrix} x \\ y \end{bmatrix} \tag{1.11}$$

This equation tells us about the required factors associated with each of the components (x, y) when a vector is rotated by an angle θ. The inverse transformation tells if the primed components are known, how we can figure out where they have rotated from by an angle $-\theta$. Then we need to simply find an inverse rotation matrix, which when multiplied by the original matrix gives the identity matrix.

$$\begin{bmatrix} x \\ y \end{bmatrix} = \begin{bmatrix} \cos\theta' & -\sin\theta' \\ \sin\theta' & \cos\theta' \end{bmatrix} \begin{bmatrix} x' \\ y' \end{bmatrix}$$

$$= \begin{bmatrix} \cos\theta & -\sin\theta \\ \sin\theta & \cos\theta \end{bmatrix} \begin{bmatrix} \cos\theta & \sin\theta \\ -\sin\theta & \cos\theta \end{bmatrix} \begin{bmatrix} x \\ y \end{bmatrix}$$

$$= \begin{bmatrix} 1 & 0 \\ 0 & 1 \end{bmatrix} \begin{bmatrix} x \\ y \end{bmatrix} = \begin{bmatrix} x \\ y \end{bmatrix}$$

So the inverse transformation matrix is the matrix of the same form but it takes the vector back to the original reference frame. Another way to describe the inverse transformation is that the basis vectors are inverted for that purpose. A transformation which gives an identity by multiplying with the inverse transformation is called a unitary transformation.

When a vector is transformed from one vector space to another vector space preserving its magnitude, the transformation is called a covariant transformation, whereas the inverse transformation that can take back the transformed vector to the original vector space is called a contravariant transformation.

Tensors exhibit the transformation properties of different quantities associated with the inherited transformation of vector space in terms of transformation of its basis vectors corresponding to two vector spaces. A matrix is composed of all the elements represented by c-numbers where a tensor has the information based on how each and every individual basis vector transforms as a linear combination of all the basis vectors of the transformed vector space.

The rotation of a three-dimensional vector transformed in 2D space around a fixed axis, for all three components x_j, for $i = 1$, 2 and 3 can be written as

$$\begin{aligned} x'_1 &= x_1\cos(\theta) + x_2\sin(\theta) \\ x'_2 &= -x_1\sin(\theta) + x_2\cos(\theta) \\ x'_3 &= x_3 \end{aligned} \tag{1.12}$$

The rotation of the xy-plane about the z-axis in matrix form can be written as:

$$\begin{pmatrix} x'_1 \\ x'_2 \\ x'_3 \end{pmatrix} = \begin{pmatrix} \cos(\theta) & \sin(\theta) & 0 \\ -\sin(\theta) & \cos(\theta) & 0 \\ 0 & 0 & 1 \end{pmatrix} \begin{pmatrix} x_1 \\ x_2 \\ x_3 \end{pmatrix}$$

and the most general form of three-dimensional matrix transformation is:

$$\begin{pmatrix} x'_1 \\ x'_2 \\ x'_3 \end{pmatrix} = \begin{pmatrix} a_{11} & a_{12} & a_{13} \\ a_{21} & a_{22} & a_{23} \\ a_{31} & a_{32} & a_{33} \end{pmatrix} \begin{pmatrix} x_1 \\ x_2 \\ x_3 \end{pmatrix}$$

We can say that $\mathbf{x}'$ is a transformed vector in primed coordinates that has been obtained by the rotation of the same vector from the un-primed coordinate space as $\mathbf{x}$, represented in tensor notation in a compact form as:

$$x'_l = \sum_m a_l^{\ m} x_m = a_l^{\ m} x_m \text{ (summation is understood)}$$

and reads as the transformation of vector $\mathbf{x}$ into the vector $\mathbf{x}'$ under rotation around an axis. The rotation matrix a_l^m and one of its examples is in equation (1.11) as a rotation about the z-axis by an angle θ. The transformation matrices a_l^m are always square matrices. A dummy index like 'm' in the above equation is always summed whenever it appears in the superscript and the subscript of two terms in a product. The dummy indices do not need to be next to each other. A dummy index is a repeated index that cannot be used more than once in the same term. Therefore, a dummy index can always be changed without affecting the results. For example,

$$a_m^n = c_m^i c_i^j c_j^n = c_m^r c_r^s c_s^n = c_m^r c_r^s c_s^n.$$

Dummy indices are summed up and cancel each other and they do not appear in the reult. They are repeated as contravariant and covariant index and cancel each other. A successive transformation of the same vector $\mathbf{x}$ into $\mathbf{x}''$ is then given by

$$x''_l = \begin{pmatrix} a'_{11} & a'_{12} & a'_{13} \\ a'_{21} & a'_{22} & a'_{23} \\ a'_{31} & a'_{32} & a'_{33} \end{pmatrix} \begin{pmatrix} a_{11} & a_{12} & a_{13} \\ a_{21} & a_{22} & a_{23} \\ a_{31} & a_{32} & a_{33} \end{pmatrix} \begin{pmatrix} x_1 \\ x_2 \\ x_3 \end{pmatrix} \tag{1.13}$$

In tensor notation, it is written as:

$$x''_l = a_l^{\ m} x'_m = a_l^{\ m}(a_m^n x_n) = a_l^{\ m} a_m^n x_n = b_l^{\ n} x_n$$

where,

$$b_l^{\ n} = a_l^{\ m} a_m^n,$$

and m is a dummy index and a summation over m gives matrix elements of the new matrix b_l^n. So a column vector basically needs one transformation matrix. This transformation can be a single step transformation or takes in two or more steps.

However, at the end, all those transformations can be represented as a single transformation given by b_l^n in the above equation.

This process can be continued through identifying tensors by their rank. The rank can be defined by how many independent transformation matrices (with no dummy indiex) are needed to transform. The rank of matrices can be easily linked by the transformation indices associated with tensors such that we can consider a scalar quantity as a tensor of rank 0 as it is not described in terms of its components in the form of basis vectors in a vector space. The number of rows and columns of the transformation matrices depend on the number of basis vectors required to generate the given vector space.

Scalars are zero-rank tensors and are moved within the vector space without transformation of its components so they are just represented by their magnitude and not the direction. Vectors are, on the other hand, a tensor of rank 1, basically because their movements within the vector space are described in terms of a transformation matrix a_l^m and every element of the matrix element can be defined as a direction cosine of the basis vector of one vector space to the basis vectors of the other corresponding vector space.

A well-known example of a transformation matrix of a vector in Cartesian coordinates can be a rotation of a vector about one axis. All of the successive rotations can be combined together as a single rotation. We show two successive rotations of the same vector with a positive and negative angle of rotation. We can continue this process by multiplication of the transformation matrices. If we can multiply two matrices to one matrix, the rank of the tensor is reduced to 1.

$$a_l^m a_m^n = \begin{pmatrix} \cos(\theta) & \sin(\theta) & 0 \\ -\sin(\theta) & \cos(\theta) & 0 \\ 0 & 0 & 1 \end{pmatrix} \begin{pmatrix} \cos(\phi) & \sin(\phi) & 0 \\ -\sin(\phi) & \cos(\phi) & 0 \\ 0 & 0 & 1 \end{pmatrix} \tag{1.14}$$

$$= \begin{pmatrix} \cos(\theta + \phi) & \sin(\theta + \phi) & 0 \\ -\sin(\theta + \phi) & \cos(\theta + \phi) & 0 \\ 0 & 0 & 1 \end{pmatrix} \tag{1.15}$$

When $\theta + \phi = 0$ or $\theta = -\phi$ we obtain the identity rotation

$$a_l^n \equiv \delta_l^n = \begin{pmatrix} 1 & 0 & 0 \\ 0 & 1 & 0 \\ 0 & 0 & 1 \end{pmatrix} \tag{1.16}$$

The matrices a_l^m and a_m^n are two successive rotation matrices about the z-axis (x_3) by an angle θ and ϕ, respectively, such that the net rotation appears to be a sum of those rotations. The combination of these rotation angles helps to develop trigonometric identities such that the net rotation comes out to be the sum of or difference of successive rotation angles, given in equation (1.15). As a special case, we consider both rotations are by the angle of the same magnitude but different direction giving an overall identity transformation given in equation (1.16).

The identity matrix in any vector space with any number of basis vectors has all the diagonal elements as unity and off-diagonal elements as zeros. It is also worth mentioning that the transformation matrices have to be square matrices as they transform vectors from one to the other coordinate space without changing its dimensions. The same rule follows for tensors also.

1.2.4 Transformation of coordinates

Transformation matrices help to relate various coordinate systems. The most comonly used three-dimensional coordinate systems are Cartesian coordinates (X) and the spherical polar coordinates (R). It is therefore important to know the transformation between the Cartesian and spherical polar coordinates. Spherical coordinates get special importance due to the spherical nature of both long distance fundamental forces, namely gravity and electromagnetic interactions. Both of them follow the inverse square law of force and are represented by radial potential. The third commonly used coordinate system is the cylindrical coordinate system (P). In this section we will express transformation equations for these coordinate systems and we will use them to solve differential equations later. We can express x, y, z coordinates in terms of spherical polar coordinates r, θ and ϕ for $r^2 = x^2 + y^2 + z^2$ as:

$$x = r \sin\theta \cos\phi$$

$$y = r \sin\theta \sin\phi$$

$$z = r \cos\theta$$

where θ is the polar angle and ϕ is the azimuthal angle. The three-dimensional coordinate transformation can be written in the matrix form of direction cosines expressed as:

$$\begin{bmatrix} dx \\ dy \\ dz \end{bmatrix} = \begin{bmatrix} \sin\theta\cos\phi & r\cos\theta\cos\phi & -r\sin\theta\sin\phi \\ \sin\theta\sin\phi & r\cos\theta\sin\phi & -r\cos\theta\cos\phi \\ \cos\theta & -r\sin\theta & 0 \end{bmatrix} \begin{bmatrix} dr \\ d\theta \\ d\phi \end{bmatrix} \tag{1.17}$$

$$dX_i = h_i^j dR_j \tag{1.18}$$

$$dX = HdR \tag{1.19}$$

and the small volume element transformation used for the transformation of integration variable between two coordinate systems can be written as:

$$dxdydz = \det Hdrd\theta d\phi = r^2 \sin\theta \cos\phi drd\theta d\phi \tag{1.20}$$

The cylindrical coordinates P are produced by a generalization of plane polar coordinates such that we can express x, y, z coordinates in terms of cylindrical coordinates ρ, θ and the length of a cylinder z for $\rho^2 = x^2 + y^2$ as:

$$x = \rho \cos\theta$$

$$y = \rho \sin\theta$$

$$z = z$$

where θ is the polar angle and ϕ is the azimuthal angle. The three-dimensional coordinate transformation can be written in the matrix form of direction cosines is written as

$$\begin{bmatrix} dx \\ dy \\ dz \end{bmatrix} = \begin{bmatrix} \cos\theta & -\rho\sin\theta & 0 \\ \sin\theta & \rho\cos\theta & -r\cos\theta \\ 0 & 0 & 1 \end{bmatrix} \begin{bmatrix} d\rho \\ d\theta \\ dz \end{bmatrix} \tag{1.21}$$

$$dX_i = M_i^j dP_j \tag{1.22}$$

$$dX = MdP \tag{1.23}$$

and the small volume element transformation used for the transformation of integration variables between two coordinate systems can be written as:

$$\begin{aligned} dxdydz &= \det H dr d\theta d\phi \\ &= r^2 \sin\theta \cos\phi dr d\theta d\phi \end{aligned} \tag{1.24}$$

and the volume element is given by

$$\begin{aligned} dxdydz &= \det M d\rho d\theta dz \\ &= \rho d\rho d\theta dz. \end{aligned} \tag{1.25}$$

1.3 Generalization of matrices into tensors

Matrices are generalized to tensors. Matrices, as two-dimensional arrays, can be expressed on paper and the higher-dimensional algebra is developed as a generalization for matrices as a part of linear algebra. However, three- or higher-dimensional arrays cannot be described on a piece of paper. So a generalized form of higher-dimensional algebra is required to accommodate higher-dimensional mathematics. Tensor algebra can be considered a straightforward generalization of the matrix algebra. Vector algebra and matrix algebra can be derived from tensor algebra.

Vectors of tensors of rank 1 can be written as a single array. Matrices are tensors of rank 2 because they are two-dimensional arrays which can vary by two indices, one along the horizontal side and the other along the vertical direction. Every element in the matrix is identified by a corresponding row and column. Rotation of vectors can be described by one transformation matrix equation (1.11) or as a rotation, expressed by the transformation matrix of rotation, given by equation (1.14). This transformation can be expressed in the compact form by equation (1.15). Second-rank tensors need two transformation matrices to transform each index and equation (1.15) can be generalized as:

$$T'_{lm} = \sum_i \sum_j a_l^i a_m^j T_{ij} = \sum_{i,j} a_l^i a_m^j T_{ij} \tag{1.26}$$

where, $i, j, \ldots, k$ is a set of indices in one vector space and $l, m, \ldots, n$ are in the other vector space, all are three-dimensional indices. We can generalize these indices to any dimensions and the rank of a vector can be determined independently. In the most general form, we can write a tensor of rank N in this form:

$$\begin{aligned} T'_{i,j,\ldots,k} &= \sum_l \sum_m \cdots \sum_n a_i^l a_j^m \cdots a_k^n T_{lm\ldots n} \\ &= \sum_{l,m,\ldots,n} a_i^l a_j^m \cdots a_k^n T_{lm\ldots n} \end{aligned} \tag{1.27}$$

Tensors involve summation of dummy indices, which actually correspond to multiplication of matrices and cause the reduction of the rank of a tensor. Matrix multiplication is only possible if the number of rows of one matrix matches with the number of columns of the multiplying matrix. It can be seen from linear algebra as $M_{lm} \times M_{mn} = M_{ln}$. In matrix form it reads:

$$\begin{pmatrix} a''_{11} & a''_{12} & a''_{13} \\ a''_{21} & a''_{22} & a''_{23} \\ a''_{31} & a''_{32} & a''_{33} \end{pmatrix} = \begin{pmatrix} a'_{11} & a'_{12} & a'_{13} \\ a'_{21} & a'_{22} & a'_{23} \\ a'_{31} & a'_{32} & a'_{33} \end{pmatrix} \begin{pmatrix} a_{11} & a_{12} & a_{13} \\ a_{21} & a_{22} & a_{23} \\ a_{31} & a_{32} & a_{33} \end{pmatrix} \begin{pmatrix} x_1 \\ x_2 \\ x_3 \end{pmatrix} \tag{1.28}$$

The transformation matrix in n-dimensional form can be generalized to any larger dimension matrix $b_l n$ as:

$$b_{ln} = \begin{pmatrix} a_{11} & a_{12} & \cdots & a_{1n} \\ a_{21} & a_{22} & & a_{2n} \\ \vdots & & & \\ \vdots & & & \\ a_{l1} & a_{l2} & & a_{ln} \end{pmatrix}$$

Transformation of matrices occurs in the same vector space or from one space to another space only if the dimensions of vector space remain unchanged. The requirement of the matching number of rows and columns of transformation vectors are obviously related to the dimension of the vector space as the index of a tensor corresponds to the number of basis vectors defining a vector space. All of the transformation matrices are always square matrices. Keeping in mind the matrix multiplication requirement, the number of columns of the first matrix should be equal to the number of rows of the second matrix and the resultant matrix has the number of rows of the first matrix and number of columns of the second matrix. In the general form, we can write:

$$b_{(\text{rows}\times\text{columns})} = a_{(\text{rows}\times n)}a_{(n\times\text{columns})}$$

A dot product provides a good example as

$$a_{(1\times 3)}a_{(3\times 1)} = \begin{pmatrix} x_1 & x_2 & x_3 \end{pmatrix}\begin{pmatrix} y_1 \\ y_2 \\ y_3 \end{pmatrix}$$

is given as a scalar or tensor of rank 0, which can just be a c-number.

$$b_{(1\times 1)} = x_1y_1 + x_2y_2 + x_3y_3 \equiv (c\text{-number})$$

However, a three-dimensional square matrix is produced as a matrix product of two vectors:

$$a_{(3\times 1)}a_{(1\times 3)} = \begin{pmatrix} y_1 \\ y_2 \\ y_3 \end{pmatrix}\begin{pmatrix} x_1 & x_2 & x_3 \end{pmatrix}$$

The multiplication of transformation matrices requires the representation of rows as lower indices and columns as upper indices to explain their ability to be able to multiply to give one matrix instead of two, such that the above multiplication can be written as:

$$b_{\text{rows}}^{\text{columns}} = a_{\text{rows}}^{n}a_{n}^{\text{columns}}, \tag{1.29}$$

and n is called the dummy (or summed) index. In the tensor notation, a summation over a dummy index is also called the contraction of indices that shows the multiplication of two matrices.

1.3.1 Tensors

Tensors are defined in terms of the transformation from one frame of reference to another. Scalars are tensors of rank 0 as they remain unchanged from one system to another. Vectors are tensors of rank 1 or a single array (row or column) as they can transform from one coordinate system to another coordinate system. One transformation matrix can only transform one array (vector) from one frame of reference to another one and is identified as a tensor of rank 1. Matrices are defined by n-dimensional arrays as a product of n-rows and n-columns (or $n \times n$ elements). Therefore, two transformation matrices are needed to transform a matrix from one frame to another and they are identified as second-rank tensors. Tensor representation becomes very useful as we can generalize them to higher ranks easily, even if we cannot write them on two-dimensional sheets. For this purpose, we generalized vector and matrix algebra to tensor algebra as well.

1.3.2 Covariant and contravariant tensors

Covariant vectors are written in terms of an array of its components or a column matrix. These vectors can be transformed from one coordinate system to another coordinate system using transformation matrices. And these transformation matrices are called tensors. Tensors give the transformation of the vector space in terms of its basis vectors. So the transformation of a vector from one coordinate system to the other systems occurs due to the transformation of basis vectors and the projection of a vector along the transformed basis vectors is changed. The complete set of basis vectors preserves the magnitude of the vector. This multiplication of two matrices to give one matrix is required for transformation matrices and is indicated by representing rows as lower indices and columns as upper indices to explain their ability to be able to multiply. Equation (1.29) provides an example. The dummy index n indicates that the multiplication is only possible if the number of columns of the first (left-hand side) matrix is equal to the number of rows of the second (right-hand side) matrix. Quantum mechanically, outgoing conjugate states are represented as row matrices and incoming states are represented as column matrices, and the product is just a number equal to the probability. The column vectors or the tensors that have the same number of rows in the dummy index are covariant tensors and the row vectors are called contravariant tensors or in three-dimensional space the original vector $\mathbf{x}$ is a covariant vector and is transformed to a rotated space as $\mathbf{x}'$ through a rotation matrix as shown in equation (1.13). The rotated vector $\mathbf{x}'$ is a contravariant vector that can be rotated back to the original vector $\mathbf{x}$, using the inverse transformation. Similarly, in a four-dimensional formalism, we can consider a vector in stationary frame as a covariant vector and in the moving frame as contravariant vector. Covariant vectors will use Lorentz transformations and contravariant vectors can be transformed back to the stationary frame by inverse transformation. To identify covariant vectors, we wrote equation (1.27) as:

$$T'_{i,j,\ldots,k} = \sum_{l,m,\ldots,n} a_i^l a_j^m \cdots a_k^n T_{lm\ldots n}$$

where $i \to j \to k$ changes counterclockwise for the antisymmetric tensor ε_{ijk}. It is symmetric under the change of sign counterclockwise and it creates an extra negative sign for a clockwise change of indices. The benefit of tensor notation is that we can write any higher-order tensors in a given space. The variation of indices indicates the dimensionality as $i, j, k, \ldots$ vary from 1 to 3 in three-dimensional space and 1–4 (or sometimes 0–3) in four-dimensional space. Even the n-dimensional generalization is straightforward making the variation of indices from 1 to n. However, just for convenience, Latin indices are used for two and three-dimensional indices and Greek indices are used for four-dimensional coordinates. Another convenience lies in the fact that the order of their operation and multiplicity is indicated by the repeated indices and not how the matrices are written. Covariant transformation does not change the magnitude of a vector even if the basis vectors are transformed. A well-known example is the transformation of a scalar from one coordinate system to another one without changing its magnitude. If a vector is brought from an outside vector space to the previous vector space, it transforms using contravariant transformation, that is the inverse of the covariant transformation and can be written as:

$$\frac{\partial f}{\partial x^i} = \frac{\partial f}{\partial x'^l}\frac{\partial x'^l}{\partial x^i} \tag{1.30}$$

l is summed over three coordinates, being a dummy index, such that the transformation of basis vector can be given as

$$e'_i = \frac{\partial x^l}{\partial x'^i} e_l \tag{1.31}$$

and the corresponding coordinate transformation can be written as:

$$dx'^i = \frac{\partial x'^i}{\partial x^l} dx^l \tag{1.32}$$

The covariant and contravariant transformations are the same in three-dimensional coordinates. However, we can still define the contravariant transformation as:

$$dx^i = \frac{\partial x^i}{\partial x'^l} dx'^l \tag{1.33}$$

Covariant transformation of matrices, as second-rank tensors, can be written as:

$$A'_{ij} = \frac{\partial x^l}{\partial x'^i}\frac{\partial x^m}{\partial x'^j} A_{lm} \tag{1.34}$$

The corresponding inverse transformation, called a contravariant transformation of matrices, can be described as:

$$A'^{ij} = \frac{\partial x'^i}{\partial x^l}\frac{\partial x'^j}{\partial x'^m} A^{lm} \tag{1.35}$$

If there is a combination of covariant and contravariant transformation, it could be written as:

$$A_j'^i = \frac{\partial x'^i}{\partial x^l}\frac{\partial x^m}{\partial x'^j} A_m^l \tag{1.36}$$

If the mixed covariant and contravariant transformation reduces the rank of a tensor by two, it is the inverse transformation. In this case we can write

$$\delta'^i_j = \frac{\partial x'^i}{\partial x^l}\frac{\partial x^l}{\partial x'^j} \tag{1.37}$$

and A is the measurable quantity, which is physically observable. Kronecker delta in equation (1.37) insures the contraction of tensors and helps to remove a dummy index, changing it into an identity matrix. A will be identified as an invariant quantity. The product of two vectors can generate a tensor of rank 2 or a tensor lower than rank 2, if the product is taken in the same vector space. Such a product is called a reducible tensor, and can then be written in terms of the symmetric and antisymmetric product of two vectors of equation (1.3) and can then be written as three possible terms of vector multiplication as

$$\mathbf{AB} \rightarrow \mathbf{A} \cdot \mathbf{B} + \mathbf{A} \times \mathbf{B} + \mathbf{M}$$
$$A_i B_j = \sum_j \delta_{ij} A_i B^j + \sum_{j \neq k} \varepsilon_{ijk} A^j B^k + A^i B_j$$

which can then be represented as:

$$\begin{bmatrix} A_1 \\ A_2 \\ A_3 \end{bmatrix} \begin{bmatrix} B_1 \\ B_2 \\ B_3 \end{bmatrix} \longrightarrow \begin{bmatrix} A_1 & A_2 & A3 \end{bmatrix} \begin{bmatrix} B_1 \\ B_2 \\ B3 \end{bmatrix} \begin{pmatrix} 1 & 0 & 0 \\ 0 & 1 & 0 \\ 0 & 0 & 1 \end{pmatrix}$$
$$+ \left((A_2 B_3 - B_2 A_3) \begin{bmatrix} 1 \\ 0 \\ 0 \end{bmatrix} + (A_3 B_1 - A_1 B_3) \begin{bmatrix} 0 \\ 1 \\ 0 \end{bmatrix} + (A_1 B_2 - A_2 B_1) \begin{bmatrix} 0 \\ 0 \\ 1 \end{bmatrix} \right) +$$
$$\begin{bmatrix} A_1 B_1 & A_1 B_2 & A_1 B_3 \\ A_2 B_1 & A_2 B_2 & A_2 B_3 \\ A_3 B_1 & A_3 B_2 & A_3 B_3 \end{bmatrix}$$

which is also called a reducible tensor as it can produce a sum of all possible tensors of lower-rank terms. Here we have all of i, j and k indices in three-dimensional space. The difference between covariant and contravariant indices in three-dimensional space is not possible, so all of the three-dimensional transformations are covariant (real space) and directly related to each other by very simple transformations. The four-dimensional space clearly distinguishes between the covariant and contravariant transformations and the difference between them is clearly specified as the measurable changes can be seen with respect to three-dimensional space.

1.3.3 Four-dimensional tensors

The four-dimensional space represents the coordinate space in relativity and instead of measurable length of an object, it can describe an event in the physical sense. The four-dimensional analogue of length is indicated as spacial length at a given time. Four-dimensional tensors or four-vectors are expressed by three spatial coordinates and time as the fourth coordinate represented as

$$(x, y, z, ict) = (x_1, x_2, x_3, x_4)$$

where space is real and time is an imaginary coordinate such that the fourth coordinate $x_4 = \iota ct$. Four-dimensional tensor indices are usually Greek indices such as (α, β, μ, v,) etc. So the four-vector transformation is given as $x^\mu \equiv (ct, x, y, z)$ and μ runs from 1 to 4 as four-dimensional transformation, x'_μ, which is given by:

$$x'_\mu = \sum_{v=0}^{3} a_\mu^v x_v \tag{1.38}$$

$$a_\mu^v = \begin{pmatrix} \frac{1}{\sqrt{1-\frac{v^2}{c^2}}} & \frac{-v/c}{\sqrt{1-\frac{v^2}{c^2}}} & 0 & 0 \\ \frac{-v/c}{\sqrt{1-\frac{v^2}{c^2}}} & \frac{1}{\sqrt{1-\frac{v^2}{c^2}}} & 0 & 0 \\ 0 & 0 & 1 & 0 \\ 0 & 0 & 0 & 1 \end{pmatrix} \tag{1.39}$$

Since $x^\mu x_\mu = x^2 = x_0^2 - x_1^2 - x_2^2 - x_3^2$ is defined as the magnitude in four-dimensional space we can write an infinitesimal transformation $dx^{'2}$ as $dx^{'2} = \eta_{\mu v} dx_\mu dx_v$ with the metric tensor $\eta_{\mu v}$ in Minksowski space for real time given by:

$$\eta_{\mu v} = \begin{pmatrix} 1 & 0 & 0 & 0 \\ 0 & -1 & 0 & 0 \\ 0 & 0 & -1 & 0 \\ 0 & 0 & 0 & -1 \end{pmatrix} \tag{1.40}$$

To develop an intuition for covariant transformations, we can consider that a covariant transformation is a transfer of information from the observer's frame to a distant frame and the inverse transformation is called the contravariant transformation and is responsible for transformation of the event from a distant frame to the frame of observer. We can show that the covariance requires that the transformation matrices are the inverse of each other for the covariant and contravariant transformations and the above matrix is given as:

$$\eta'_{\mu v} = a_\mu^\alpha \eta_{\alpha\beta} a_v^\beta .$$

Now consider the inverse transformation, and using $x^{'\alpha} = a_\mu^\alpha x^\mu$, we can write

$$\eta'_{\alpha\beta} x^{'\beta} x^{'\alpha} = \eta_{\alpha\beta} a_\mu^\beta x^{'\mu} a_v^\alpha x^{'v} = \eta_{\alpha\beta} a_\mu^\beta a_v^\alpha x^{'\mu} x^{'v}$$

such that we rewrite the above relations as:

$$\eta'_{\alpha\beta} x^{'\beta} x^{'\alpha} = \eta_{\alpha\beta} x^\alpha x^\beta$$

or plugging in the values of inverse transformation, we can show that:

$$\eta_{\alpha\beta}(a_\mu^\beta a_v^\alpha) x^{'\mu} x^{'v} = \eta_{\mu v} x^\mu x^v$$

Subtracting the right-hand term from both sides:

$$(a_\mu^\beta \eta_{\alpha\beta} a_v^\alpha - \eta_{\mu v}) x^\mu x^v = 0$$

This leaves us only with

$$a_\mu^\beta \eta_{\alpha\beta} a_v^\alpha - \eta_{\mu v} = 0$$

and finally the invariance of the transformation matrix in the same vector space can be shown as:

$$\boxed{\eta_{\mu v} = a_\mu^\beta \eta_{\alpha\beta} a_v^\alpha}$$

All of the vector identities can be expressed in the convenient way in the tensor notation and the vector algebra can be performed in a much more convenient way in the tensor notation. The del operator is simply treated as a vector with its covariant and contravariant components by transforming $\nabla_i \rightarrow \partial_\mu$ in four-dimensional space. Moreover, four-dimensional vectors are not represented as bold face vectors. Therefore, we express a few useful vector indentities in tensor notations, giving divergence and curl operators in the same notation as given below.

$$\begin{aligned} A \cdot B &= \sum_i A_i B^i \\ \nabla \cdot B &= \sum_i \partial_i B^i \\ (A \times B)_i &= \varepsilon_{ijk} A^j B^k \\ (\nabla \times B)_i &= \varepsilon_{ijk} \partial^j B^k \end{aligned} \tag{1.41}$$

All of the vector identities and then vector operations in three-dimensional space can also be obtained using three-dimensional letters as tensor indices instead of Greek letters as four-dimensional indices. This is how conveniently a complete set of three-dimensional vector identities can be transformed into four-dimensional identities or even transformations to different vector spaces can be expressed using tensors. It is not only a compact notation, it simplifies the vector calculus more conveniently by tremendous simplifications in its cumbersome calculations. It happens because we do not need to write complete vector terms and simply use indices plus the order of operation is easy to maintain due to the assigned indices. The covariant and contravariant tensors are practically associated with the reference point. Therefore, within the same vector space, contravariant and covariant transformations are almost similar so the mixing of upper and lower indices does not matter in three-dimensional vector space. Tensors give a general description of vectors and their products that lead to measurable quantities in physics and are used as an excellent mathematical description of complicated quantities.

We live in three-dimensional space and all lab measurements are made in space at certain times, so classical physics is studied in three-dimensional flat space. However, special relativity includes time as an imaginary coordinate to understand relative motion of fast-moving objects. Special relativity provides an excellent tool in high-energy physics and quantum field theory is developed incorporating the relativistic motion of individual particles such that the lab measurements can be translated into the individual particle properties using the appropriate frame of reference.

All of the physical measurements are only a part of the full information, and they are possible in four-dimensional space only. Most quantities in physics can be written as a combination of two vectors with a dot or cross product. Scalar products represent the physical parameters that are generated by a combination (e.g., dot product) of two vectors such as area that is generated by the projection of one vector along another vector. A cross product gives a pseudo-vector and both of them can be expressed in terms of tensors very well. The products of vectors are related to their dimensionality such that a two-dimensional vector can give a scalar or a vector.

1.3.4 Application of tensors

Tensor relations are the same expressions that describe the algebra of scalars, vectors, matrices, and multiple arrays all at one place in a very compact form differentiating them by rank only. Scalars are tensors of rank 0, vectors have rank 1, and matrices are tensors of rank 2. However, tensors of rank 3 and more can only be written in mathematical form. The most well-known tensors are associated with coordinate space. The simplest vectors in three-dimensional space can be transformed from one coordinate system to another one using direction cosines. Another example of transformation matrices is a rotation matrix that transforms a spatial vector from an initial state to a final state.

Tensors are effective tools to describe dynamics of many-body systems where individual particle properties have to be integrated together and multiple parameters affect differently different particles. It then becomes possible to express the transformation of a state of the system into another state. A many-particle system composed of n particles can be considered as an array of n components. This system is initially giving an n-dimensional vector and its final state is another vector in n-dimensional state. Inter-particle interactions among these particles are identified as a $(n \times n)$ transformation matrix. n then corresponds to the dimensionality of a tensor.

Each index of a tensor is transformed from the initial state to the final state of the system by multiplying with a transformation matrix. The number of transformation matrices needed to change the state of a system is called the rank of a tensor. In other words, the rank of a tensor is defined by the number of indices associated with a tensor. Each index corresponds to an array in one direction. As a rank 2 tensor, matrices correspond to a combination of initial and final states. A transformation matrix indicates how to transform a column of initial state into a row of final state or vice versa. This is what is related to contraction of an index.

The contraction of tensors by the presence of dummy indices leads to the reduction of their rank, whereas their multiplication can lead to increase or decrease in their rank depending on how the multiplication is done. Tensor calculus describes the variation of tensors in four-dimensional space and geometric algebra takes care of the tensor operations in its own parameters. The rank of a tensor cannot be greater than its dimensionality. Therefore, we can only express a tensor up to rank 2 (matrix) in two-dimensional space. Since we can write a matrix in block representation, this can help to express the four-dimensional space as a two-dimensional space–time sheet.

A connection between matter and energy can be established in Euclidean geometry in four-dimensional coordinate space that relates space and time together. This four-dimensional representation is the only way to describe the energy–momentum as interchangeable coordinates and show the relationship between electric and magnetic fields via Maxwell's equation. The conjugate variables of position and momentum are related to the uncertainty principle or the concept of wave–particle duality in quantum mechanics. Similarly, energy and time are shown to be conjugate variables just as momentum and spacial coordinates can be related as conjugate variables. These conjugate variables relate to quantum mechanics and help to develop a powerful mathematical technique of Fourier transformation that is a very powerful mathematical technique for electrical engineers.

All of the physical measurements are only a part of the full information, and they are possible in four-dimensional space only. Most quantities in physics can be written as a combination of two vectors with a dot or cross product. Scalar products represent the physical parameters that are generated by a combination (e.g., dot product) of two vectors such as area that is generated by the projection of one vector along another vector. A cross product gives a pseudo-vector and both of them can be expressed in terms of tensors very well. The products of vectors are related to their dimensionality such that a two-dimensional vector can give a scalar or a vector.

1.4 Geometric algebra

Geometric algebra provides a tool to describe all kinds of combinations of vectors and defines multivectors instead of giving dot and cross product. Higher combinations of vectors can be defined in this way as well. In three-dimensional space, we define tensors of rank 0 as scalars, rank 1 as vectors and rank 2 as matrices. In four-dimensional space, tensors of rank 3 can also be defined. So the rank of a tensor has to be lower than the dimensionality of a space to be expressible in that space.

Three-dimensional space with real spatial components is known as Euclidean space and the inner product of two vectors in this vector space is defined as $(\mathbf{a},\mathbf{b}) = \mathbf{a}\cdot\mathbf{b}$. The symmetric part of the geometric inner product is written as:

$$\mathbf{a}\cdot\mathbf{b} = \frac{1}{2}(\mathbf{ab} + \mathbf{ba}) = \frac{1}{2}[(\mathbf{a} + \mathbf{b})^2 - \mathbf{a}^2 - \mathbf{b}^2] \tag{1.42}$$

Inner product satisfies commutative and associative laws and is linear and reduces the rank of combined vectors by one or contracting the same index of tensors. So the dot product produces a zero rank tensor. It can be considered as area swiped by a vector to project on the other vector. This area is the same if vector 1 approaches vector 2 or 2 approaches 1 and is a scalar quantity. Its orientation does not matter.

The outer product of two vectors, on the other hand, maintains the same rank and therefore can even be defined in two-dimensional space. It is an antisymmetric combination and gives a so called pseudo-vector instead of a vector and is then defined as:

$$\mathbf{a} \wedge \mathbf{b} = \frac{1}{2}(\mathbf{ab} - \mathbf{ba}) = \frac{1}{2}[(\mathbf{a} - \mathbf{b})(\mathbf{b} - \mathbf{a}) + \mathbf{a}^2 + \mathbf{b}^2] \tag{1.43}$$

Thus, the geometric product of two vectors (bi-vectors) is a mathematical quantity that describes the antisymmetric combination of two vectors and is defined in exterior algebra or geometric algebra. At this stage, we define outer product as:

$$\mathbf{ab} = \mathbf{a} \cdot \mathbf{b} + \mathbf{a} \wedge \mathbf{b} = \frac{1}{2}(\mathbf{ab} + \mathbf{ba}) + \frac{1}{2}(\mathbf{ab} - \mathbf{ba}) \tag{1.44}$$

In the component form in three-dimensional space, a bivector is completely defined as:

$$\mathbf{ab} = \mathbf{a}_1 b_1 - a_2 b_2 + c_1 c_2 + (a_1 b_2 - b_1 a_2)\mathbf{e}_1\mathbf{e}_2 + (\mathbf{b}_1\mathbf{c}_2 - \mathbf{c}_1\mathbf{b}_2)\mathbf{e}_2\mathbf{e}_3 + (\mathbf{a}_1\mathbf{c}_2 - \mathbf{c}_1\mathbf{a}_2)\mathbf{e}_1\mathbf{e}_3 \tag{1.45}$$

Now we already know that we can write the product of two basis vectors in three-dimensional space as:

$$\mathbf{e}_\mathbf{i} \cdot \mathbf{e}_\mathbf{j} = \delta_{\mathbf{ij}} + \varepsilon_{\mathbf{ijk}}\mathbf{e}_\mathbf{k} \tag{1.46}$$

This is exactly equivalent to the definition of bivector for basis vectors in 3D space, such that

$$\mathbf{a.b} = a_1 b_1 - a_2 b_2 + c_1 c_2 + (a_1 b_2 - b_1 a_2) \tag{1.47}$$

and

$$\mathbf{a} \wedge \mathbf{b} = (b_1 c_2 - c_1 b_2)\mathbf{e}_2\mathbf{e}_3 + (\mathbf{a}_1\mathbf{c}_2 - \mathbf{c}_1\mathbf{a}_2)\mathbf{e}_1\mathbf{e}_3 \tag{1.48}$$

and that is exactly equal to the cross product in three-dimensional space.

If the bi-vectors are required to satisfy associative law:

$$\mathbf{a} \wedge (\mathbf{b} \wedge \mathbf{c}) = (\mathbf{a} \wedge \mathbf{b}) \wedge \mathbf{c} = \frac{\mathbf{1}}{\mathbf{2}}(\mathbf{abc} - \mathbf{cba}) \tag{1.49}$$

The outer product of three vectors is called a trivector. The concept of inner product and outer product is much more used in higher dimensions. In three dimensions, wedge product and cross product simply mean the same thing. Geometric algebra is much more needed for higher dimensions since the applied physics are related to the measurement in Euclidean geometry.

1.5 Group theory

Finally, we introduce groups which provide useful tools to understand interaction theory in relativistic quantum mechanics. Groups are a special type of sets which provide a very useful tool to describe some physical problems in physics. We just give a very brief introduction of the topic to familiarize the reader to be able to fully apply it to physical problems. Otherwise, group theory in itself is a complicated subject and it requires a detailed independent study of the topic in itself.

A group is a complete set of elements which is a closed set under a (certain) binary operation. Group theory in itself is a very specialized topic in mathematics and provides an efficient tool when we need to study the physics of extended dimensions or the many-particle systems. Group theory is not applied to physics. It is extensively used in several other fields of study. We are not discussing group theory as a topic of mathematics but, just for completion, we give a brief description of a group as the interaction theories can be understood without using the representation of groups, especially the unitary groups. We will therefore include the definition of groups and briefly introduce the representation of unitary groups only.

A group is a complete set of elements which is defined under a particular binary operation and satisfies the following properties:

1. **Closure property:** This is a property which states that each and every binary operation between two elements of a group produces an element of the same group. If * is the binary operation of a group then

$$A^*B = C$$

   such that C is always a member of the same group.

2. **Associative property:** This states that the multiple of an element with a sum of two elements is equal to the sum of the product of that element with individual multiplication of two sums such that three elements (A, B, C) of a group satisfy the following relation

$$A^*(B + C) = A^*B + A^*C$$

3. **Identity element:** The set has to have an identity element e for a defined binary operation such that the application of binary operation of a group between the identity element and any other element of the group gives back the same element of the group

$$A^*e = A$$

4. **Inverse:** In a group every element of the group has an inverse of another element such that the binary operation between the elements and the inverse gives the identity element, such that

$$A^*A^{-1} = e = A^{-1*}A$$

Unitary groups are special type of group which are used to describe interaction theories and provide frameworks for the extension of coordinate space along with the scalar particles themselves. Unitary groups are the groups that include all the elements which consist of all the square matrices and a product of these matrices with its own conjugate matrix results in an identity matrix. Therefore, if we have M matrices as the elements of these groups then

$$M^{\dagger}M = MM^{-1} = I$$

where I is the identity element such that the determinant gives unity

$$det[M^{\dagger}M = MM^{-1}] = 1$$

Rotation matrices provide a good example of unitary matrices as they give a unit determinant if multiplied with its inverse. Unitary matrices give different angles of rotations if we can rotate a system by a different angle. Rotation is physically facilitated by the angular momentum, so angular momentum is called the generator of the rotation group. The generator is an operation which can produce various elements of a set. A generator uses all possible operations (by an angle θ) which generate a complete set of elements which make a group. If this set satisfies the conditions of a group, then the generator of a group describes the useful properties of a group with a great physical significance.

A detailed discussion of unification of forces using group theory, the associated extended dimensions and the corresponding particle sector is relatively more technical and group theory background is required. A discussion of the representation of groups is out of the scope of this book. Any upper-level graduate book on high-energy physics can be used to study the unification of gauge theories and the standard model.

IOP Publishing

Conceptual Approach to Quantum Electrodynamics

Samina S Masood

Chapter 2

Differential equations and the Lagrangian formalism

2.1 Differential equations

The dynamics of a physical system are associated with the applied force and conditions under which the force is applied. Response of the system towards the applied force then depends on the nature of the force and the structure and properties of the system. Dynamical behavior is described in terms of the change in properties of a system in the presence of a potential or in response to an applied force. For this purpose, symmetries and conservation rules of the interaction are incorporated to reduce the number of unknowns.

Classical physics mainly deals with the overall behavior of a system in space with time under the presence of an applied force like gravity and electromagnetic interaction, whereas quantum mechanics deals with tiny systems and is used to study the detailed structure of a system at the atomic and nuclear level. Relativity is applied to the study of mechanics of spatial objects from large distances at cosmic scales. The combination of relativity and quantum mechanics can be used to develop quantum field theory (QFT), which can be applied to study individual particle dynamics including processes at the subatomic and sub-nuclear level. Fundamental interactions play an important role in this. In short, the dynamics of everything is expressed in terms of differential equations and their solution can give the path of the underlying mechanism.

Physical processes take place due to the effect of a force produced due to interaction with other objects (particles) and is expressed in terms of differential equations which tell you the behavior of the system at every measurable moment of time and a solution of this equation can tell the trajectory of motion. Solution of these differential equations describes the dynamics of a physical system. Integration of the differential equations gives the state of a system at any given instant of time and helps to determine the trajectory of motion of the system. Understanding of the

doi:10.1088/978-0-7503-6054-8ch2

dynamics of a system depends on the proper solution of the equation of motion. Therefore, we devote the rest of this chapter to summarizing various methods of solving differential equations in reference to the particular form of equations.

Second-order differential equations (SDEs) such as Newton's laws of motion in classical mechanics, the Poisson and Laplace equations of electrodynamics and Schrödinger's equation in quantum mechanics are good examples of equations of motion of the relevant systems and the nature of the corresponding interactions. It is therefore important to briefly review basic techniques of solving differential equations. Some of the standard techniques for solving standard differential equations describe the dynamics of their systems.

2.1.1 Linear differential equations with constant coefficients

A linear differential equation is an equation comprised of a linear combination of various orders of differential operators of a coordinate that operates on a function of the same variable. A standard linear differential equation of x related to a function $f(x)$ is a polynomial equation of a differential operator D, defined as $D \equiv \frac{d}{dx}$ in powers of n such that, $D^n = \frac{d^n}{dx^n}$, is written as:

$$(a_0 + a_1 D(x) + a_2 D^2(x) + \cdots + a_n D^n(x))y(x) \equiv Ly(x) = f(x)$$

Alternatively, we can rewrite the above equation as:

$$Ly(x) = f(x)$$

and

$$L \equiv \left(a_0 + a_1\frac{d}{dx} + a_2\frac{d^2}{dx^2}\right) + a_3\frac{d^3}{dx^3} + \cdots + a_n\frac{d^n}{dx^n}. \tag{2.1}$$

The highest powers of D in the above equations, given by the index 'n', defines the order of differentiation of the equation. The number of possible solutions of a differential equation depend on the value of n, which is the highest-order derivative in this equation. Solutions of first-order differential equations (FDEs) are possible by usual methods of integration. We do not discuss higher-order equations here. However, higher-order equations can also be converted to lower-order equations and similar procedures can be followed. Every nth-order differential equation generally has n independent solutions and all of their linear combinations are also solutions of the same equation. x here could be a one-dimensional variable or could even correspond to a complete set of orthogonal variables; for example, in four-dimensional space it corresponds to a complete set of coordinates as $(x, y, z; ict)$.

A tensor representation of the most general differential equation in terms of the most general coordinate system is:

$$\sum \partial^n y(q_i) = 0 \tag{2.2}$$

where n is the order of the differential equation for finite n and q_i is a set of generalized coordinates for i number of mutually orthogonal coordinates, such that we can write:

$$y(q_i) = y(q_1)y(q_2) \cdots y(q_i) \tag{2.3}$$

The easiest way to solve a higher-order differential equation is to convert it into a lower-order equation, preferably an FDE. Quadratic equations and change of variables can lead to reducing the order of differential equations. The number of possible solutions is related to the order of the differential equation. An nth-order differential equation has n independent solutions. A linear combination of these solutions is a solution of the same equation as well. This rule can be proved by writing the Wronskian in terms of all the possible solutions to prove linear independence of various solutions. Using this rule, a SDE gives two independent solutions and the linear combinations of independent solutions can give more solutions.

The simplest form of the equation of motion of a system is obtained by taking $f(x) = 0$. In a homogeneous SDE $y(x)$ indicates the basic properties of a system. The general equation of motion is written as:

$$(a_0 + a_1D(x) + a_2D^2(x) + \cdots + a_nD^n(x))y(x) = 0 \tag{2.4}$$

which represents a situation where the operator describes the dynamics of the system without changing the fundamental properties of a system. The solution of the SDE generally describes dynamics of such a system without changing its basic properties. A higher-order differential equation typically can be converted into a lower-order equation, which can in turn be solved by using standard integrals.

SDEs are the most common equations of motion and are usually used to describe the dynamics of physical systems. The easiest method to solve a SDE is to convert it into an FDE and solve it, if possible. There are various methods to solve the FDEs. There are basically two kinds of equations, homogeneous and inhomogeneous equations. The most general homogeneous equation is then written as:

$$D^2y(x) + P(x)Dy(x) + Q(x)y(x) = 0$$

where the coefficients of the differential operator may depend on x. When it is not possible to express the equation in a complete square for derivatives or the quadratic equation may be more complicated, a series solution as an analytical function is a second option assuming that the equation is solvable. Both solutions can be provided within the series solution as even and odd functions. A general solution $y(x)$ for a SDE has two values y_1 and y_2 such that a linear combination of both solutions $y = Ay_1 + By_2$ is also a solution, for constant coefficients A and B. The functions y_1 and y_2 are obtained as solutions of the first-order equation. Conversion of a SDE to two FDEs is done using various methods including quadratic equations or making a complete square. Solutions of a first-order equation are usually found using standard integrals. In physics, sometimes a complicated first-order equation can be converted into an approximate form.

2.1.2 Second-order differential equations

We will study in detail a homogeneous SDE for:

$$Ly(x) \equiv \left(a_0 + a_1\frac{d}{dx} + a_2\frac{d^2}{dx^2}\right)y(x) = 0 \tag{2.5}$$

where a_0, a_1, a_2 represent constant coefficients. In a homogeneous SDE, a differential operator L has two roots given as:

$$L = L_1L_2,$$

which can produce two mutually independent solutions $y_1(x)$ and $y_2(x)$. These roots can be obtained by representing L as a complete square and find two first-order forms of L, such that:

$$Ly(x) = 0 = (L_1y_1)(L_2y_2).$$

Since $y_1(x)$ and $y_2(x)$ are independent of each other, the two corresponding equations are $L_1y_1(x) = 0$ and $L_2y_2(x) = 0$. Using the partial fraction method, making a complete square of L or finding roots of SDE by solving the quadratic equation can be used to find L_1 and L_2 (two independent values) to be able to easily solve the corresponding first-order equations using standard integrals. Homogeneous differential equations give a purely conserved system that obeys certain conservation rules during dynamical changes. The dynamics of such systems are related to the conservation rules of the relevant conserved forces.

2.1.3 Inhomogeneous equations

Another type of differential equations are inhomogeneous equations that cannot be converted into a first-order equation. A general inhomogeneous SDE is written as:

$$Ly(x) \equiv \left(a_0 + a_1\frac{d}{dx} + a_2\frac{d^2}{dx^2}\right)y(x) = f(x) \tag{2.6}$$

One of the simplest methods to solve an inhomogeneous SDE of a physical system is to use an analytical function as an infinite series solution that can be expressed as:

$$y(x) = \sum\nolimits_{n=0}^{\infty} c_n x^n \tag{2.7}$$

where c_n are the constant coefficients. Substitution of equation (2.7) into equation (2.6) gives a solution of the inhomogeneous as an infinite series. An infinite series solution is the most convenient method that can be used to find an analytical function. This methods gives two solutions of differential equation (2.5) as infinite series and their linear combination will be a solution as well. At this point a recurrence relation can be found among the coefficients of the infinite series given by equation (2.4). It is also worth mentioning that we have been assuming the existence of analytical solution of a differential equation to be able to write a series-solution. This approach has to be used in detail for every particular case.

Every differential equation does not have an analytical solution always, especially the inhomogeneous equation. Then various other methods can be used to solve these equations. Some of the very well-known equations can give rise to the known form of solution and special functions are constructed to solve such equations. Examples of a few well-known special functions and the relevant special equations will be discussed in the text wherever those known equations appear. These special functions are specially constructed for particular equations and are used wherever those types of equations can be written.

An inhomogeneous second-order differential equation (2.6) can also be solved in two parts. For every inhomogeneous equation one can write the homogeneous equation, that gives a solution for the region where $f(x) = 0$. We call this solution a complimentary solution written as y_C and the other inhomogeneous part can be called y_P and the total solution is written as:

$$y(x) = y_C + y_P.$$

Another method to calculate a particular solution can be found around the region where the effect of the L on the function can be maximized in terms of delta function. This method is called the Green's function method and it will be discussed later under a special topic.

Single variable equations are relatively straightforward equations, whereas the actual dynamics of a physical system is not fully described by a single variable. Even ordinary objects move in three-dimensional space and the applied force is not always parallel to the direction of motion. Therefore, the three-dimensional motion is studied in space and the change of coordinates may be needed to solve such equations.

A one-dimensional equation of motion is a totally differential equation, whereas multidimensional differential equations are expressed in terms of partial differential equations. Solutions of total differential equations or one-dimensional equations are usually possible using standard integrals tables, however. However techniques are needed to convert partial differential equations of various coordinate systems to multiple single variable (total) differential equations, which can be solved using standard integrals.

2.2 Differential equations with several variables

Movement of a physical system cannot be easily constrained in one direction only. Actual dynamics is described in space. Physical space can be considered a vector space expressed as a complete set of three basis vectors (mutually independent orthogonal coordinates) related to the movement in three-dimensional space. In this situation a differential operator becomes a vector in the coordinate space, which can be expressed as:

$$\vec{\nabla} \equiv \frac{\partial}{\partial x}\vec{i} + \frac{\partial}{\partial y}\vec{j} + \frac{\partial}{\partial z}\vec{k} \tag{2.8}$$

For mutually independent x, y and z coordinates. The partial derivatives give the opportunity to integrate the function with respect to each individual variable, assuming all other variables remain constant for that purpose. Special techniques are used to solve partial differential equations in various coordinates. We discuss below some of the commonly used techniques.

2.2.1 Partial differential equations

Partial differential equations can also be either homogeneous or inhomogeneous. Second-order partial differential equations are used to indicate the motion of an object in space; three-dimensional for non-relativistic motion and four-dimensional for relativistic motion. Homogeneous equations have a complete solution and nonhomogeneous behavior of equations of motion indicates some special behavior. The homogeneous equations of motion in the absence of any external force have standard solutions. They are relatively easy to solve as the separation of variables can be done in each coordinate and then each variable can be handled separately. So each solution can easily be expressed as a product of the independent solution in each coordinate system. The standard form of a three-dimensional second-order partial differential equation is written as:

$$[a_2 \nabla^2 + a_1 \nabla + a_0] f(x, y, z) = 0 \tag{2.9}$$

The simplest form of a SDE is obtained from $a_1 = 0$, such that:

$$[a_2 \nabla^2 + a_0] f(x, y, z) = 0 \tag{2.10}$$

and using the form of ∇^2 in various coordinates obtained by squaring equation (2.8) of chapter 1, the trajectory of motion can be studied in various coordinate systems. We use equation (2.10) as an example and the equation of motion in Cartesian coordinates can be given as:

$$\nabla^2 f(x, y, z) = \left(\frac{\partial^2}{\partial Z^2} + \frac{\partial^2}{\partial y^2} + \frac{\partial^2}{\partial z^2} \right) f(x, y, z) = 0 \tag{2.11}$$

with the solution:

$$f(x, y, z) = X(x)Y(y)Z(z) = Ae^{\pm \iota(k_x x + k_y y + k_z z)}$$

for positive values of (a_0) where the coefficients k_x, k_y and k_z are all determined from actual equations and the constant A is evaluated from the initial conditions. The positive exponential is for positive momentum and negative exponential for a negative momentum for an incoming particle.

For relativistic motion, time is added as the fourth dimension and we define the differential operator in four-dimensional $(\vec{x}, ict)$ space as

$$\left(\vec{\nabla}, \iota \frac{d}{cdt} \right) f(\vec{x}, t) \equiv \left(\frac{\partial}{\partial x} \vec{i} + \frac{\partial}{\partial y} \vec{j} + \frac{\partial}{\partial z} \vec{k} + \iota \frac{d}{cdt} \right) f(x, y, z; t) = 0 \tag{2.12}$$

and the corresponding second-order equation in four-dimensional space is written as:

$$\nabla^2 f(x, y, z) - \frac{d^2 f(x, y, z)}{c^2 dt^2} = \left(\frac{\partial^2}{\partial Z^2} + \frac{\partial^2}{\partial y^2} + \frac{\partial^2}{\partial z^2} - \frac{d^2}{c^2 dt^2}\right) f(x, y, z; t) = 0 \quad (2.13)$$

The solution of this equation will give the propagation of the wave in three-dimensional space where the time dependence corresponds to the change in phase with time and the familiar form of the four-dimensional differential equation. Using $e^{\mp \frac{E}{\hbar}t}$ as a time-dependent solution gives

$$f(x, y, z; t) \equiv X(x)Y(y)Z(z)T(t) = Ae^{\pm(k_x x + k_y y + k_z z)}e^{\mp \iota \frac{E}{\hbar}t}$$

For well-known classical wave equations $f(x, y, z; t)$ is the wavefunction and k is its solution, where k corresponds to the wave number defined as ($k = \frac{2\pi}{\lambda}$) for λ the wavelength and $\omega = \frac{E}{\hbar}$ the angular frequency, $\omega = 2\pi f$ with f as the frequency of the wave, and ω corresponds to the phase of the wave at a given time.

A solution of a differential equation in an appropriate coordinate system is more convenient than other choices. The most well-known fundamental forces are electromagnetism and gravity, which are both central forces and obey the inverse square law. For these two forces, spherical coordinates in particular are the most useful coordinates for gravity and electromagnetically-bound systems. Classical rotational motion is properly described by polar coordinates. It is always convenient to study circular motion (or rotation) in polar coordinates, whereas Cartesian coordinates are appropriate to study linear motion in flat space.

A general scheme of calculation of second-order easily solvable differential equations can be expressed as:

1. Separation of variables (if possible).
2. Express a solution in terms of a product of individual variable solutions.
3. Solve individual variable SDEs as independent equations.
4. Convert SDEs to first-order equations using the partial fraction method, making a complete square or treating derivatives as a function in the quadratic equation.
5. First-order equations are solved using the well-known standard techniques. If needed, we can define a complete solution of the equation as a linear combination of two solutions.
6. These solutions can be evaluated using initial conditions or boundary conditions.

If the SDE cannot be converted into an FDE easily, then we can write the solution of each variable (assuming an analytical solution) as an infinite series of functions of every involved single variable:

1. Write an infinite series solution of each variable.

2. Differentiate these series solutions twice and insert them back into the SDE.
3. Compare coefficients of the same power of the variable and find the relationship among various coefficients. We can find out two solutions from the same function's solutions as even or odd coefficients.
4. Sometimes, these series can be summed up and expressed as a single function.
5. A series solution can also be used to solve general differential equations with analytical solutions.

For an analytical solution of a differential equation, the Wronskian method can then be used to relate two solutions as well. If we cannot separate variables in one coordinate system, we can try another more appropriate coordinate system and separate variables. On the other hand, if a completely analytical solution is not possible, then we have to use completely different methods to solve such equations. We will discuss those solutions later. However, we need to give a quick review on multivariable equations in polar coordinates.

2.2.2 Differential equations in polar coordinates

We begin with the substitution of equation (1.4b) from chapter 1 into equation (2.10) and, after some algebra, separation of variables is possible in the following form in spherical coordinates that are obtained by squaring the differential equation in spherical polar coordinates:

$$\nabla^2 y(r,\theta,\phi) = \left(\frac{1}{r^2}\frac{\partial}{\partial r}\left(r^2\frac{\partial}{\partial r}\right) + \frac{1}{r^2\sin\theta}\frac{\partial}{\partial\theta}\left(\sin\theta\frac{\partial}{\partial\theta}\right) + \frac{1}{r^2\sin^2\theta}\frac{\partial^2}{\partial\phi^2}\right)y(r,\theta,\phi) = 0 \quad (2.14)$$

and the general solution of the above equation in spherical polar coordinates can be expressed as:

$$y(r,\theta,\phi) \equiv R(r)\Theta(\theta)\Phi(\phi)$$

where $R(r)$, $\Theta(\theta)$ and $\Phi(\phi)$ give three independent solutions. Circular motion is always an accelerated motion, and the effect of angular momentum and acceleration produced due to the centrifugal force cause the separation of the variables to be much more complicated as compared to the Cartesian coordinates in (2.13). Spherical coordinates are not easy to separate out. Special functions are then the solution of these equations which are expressed in the form of summations. Therefore, the function f can be expressed in the form of special indices n, l, m such that:

$$f \equiv R_{nl} P_l^m e^{\iota m\phi}$$

These equations are relevant to understand the orbital motions of planets in classical mechanics and are used to solve Schrödinger's equation in spherical polar coordinates. These equations, incorporating Coulomb potential, give the distribution of electrons in atoms. Since these solutions lead to very important equations, we will solve them in detail when we come across such equations in a physical system.

Solution of differential equations in cylindrical coordinates is even more complicated. Maxwell's equations indicate the existence of a electromagnetic solenoid and the motion of a charged particle in an electromagnetic field can sometimes be described well in cylindrical coordinates. In such cases, we need to use the square of equation (1.4c) from the first chapter for the same differential operator, and the equation of motion in three-dimensional space can be written in cylindrical coordinates as:

$$\nabla^2 y(\rho, \theta, z) = \left(\frac{1}{\rho}\frac{\partial}{\partial \rho}\left(\rho\frac{\partial}{\partial \rho}\right) + \frac{1}{\rho^2}\frac{\partial}{\partial \theta^2} + \frac{\partial^2}{\partial z^2}\right) y(\rho, \theta, z) = 0 \tag{2.15}$$

and:

$$y(\rho, \theta, z) \equiv P(\rho)\Theta(\theta)Z(z) \tag{2.16}$$

Other physical systems of cylindrical shapes include examples of motion of liquids in pipes or movement of current in cables, which can be described in cylindrical coordinates. However, the solution of a cylindrical equation may not always be too complicated. It may even be relatively much simpler under certain approximations. Once we learn the techniques of solving basic differential equations, we can apply those techniques to any complicated physical situation and some valid approximations for long transmission lines and long pipes may make it much simpler.

2.2.3 Nonhomogeneous partial differential equations

A general inhomogeneous equation of motion in three-dimensional space can be written as:

$$[a_2 \nabla^2 + a_1 \nabla + a_0] y(x, y, z) = f(x, y, z) \tag{2.17}$$

In the presence of an external force, an equation of motion for a system can be written in terms of a SDE. Well-known equations of motion in three-dimensional space are Newton's second law of motion, the Poisson or Laplacian equation and Schrödinger's equation for a stationary state. These common SDEs are written in the form of differential operator ∇^2, representing the kinetic energy expressed in appropriate coordinate systems. The choice of coordinates depends on the nature of the force. We mainly discuss the equations of motion related to fundamental forces. For central forces, the potential depends on the radial distance and angular dependence can be separated out as an independent variable. So the angular part of the equation may be solved by using spherical polar coordinates.

Gravity and electromagnetic forces are both radial forces and the associated potential is radial potential. Our main topic in this book is tiny objects and electromagnetic force dominates over gravity at small distances. We will therefore discuss electromagnetism as an example at small scales. In quantum electrodynamics (QED), due to the radial nature of the electromagnetic interaction, spherical coordinates in three dimensions are usually chosen to understand spatial behavior of QED in spherically symmetric potentials. Right now, we do not need to consider more complicated coordinate systems such as curvilinear coordinates because they do not have any natural global basis. All coordinate systems are mutually independent and their unit vectors are written in terms of basis vectors. The solution of such equations is expressed as a product of three independent functions f in the relevant coordinates such that:

$$f \equiv f(x, y, z) = X(x)Y(y)Z(z)$$

in Cartesian coordinates, whereas:

$$f \equiv f(r, \theta, \phi) = R(\rho), \Theta(\theta), \Phi(\phi)$$

in spherical polar coordinates and:

$$f = f(\rho, z, \phi) = P(\rho)Z(z)\Phi(\phi)$$

in cylindrical coordinates.

For a completely solvable equation, the correct form of the solution can be obtained using initial conditions of the system and the nature of the applied force. Considering a single starting equation may not be enough to find all of the unknown parameters of a system. If boundary conditions are not enough to find all unknown parameters of a system, then with the help of detailed boundary conditions, we can write more than one equation of motion for a system.

We will take an example of simple radial potential that appears in quantum mechanics and electrodynamics. The functions of the corresponding variables can be simplified and converted into a form that can be integrated easily. Then each of the definite integrals can be solved by using standard integrals. A typical example of central potentials could be:

$$\frac{d^2f}{dx^2} + \frac{d^2f}{dy^2} + \frac{d^2f}{dz^2} = \frac{f}{r} \tag{2.18}$$

In such systems, changing Cartesian coordinates into polar coordinates can help to solve such equations which may make it possible to use the separation of variables, and then we can get three independent integrals. This type of radial potential is expressed in a three-dimensional equation of spherical coordinates by integrating over the polar and the azimuthal angle as homogeneous equations. The radial part is the only part where various integration techniques can be applied for the corresponding radial potential. A general partial differential equation in spherical polar coordinates can be written as:

$$(\nabla^2 + a_1 \nabla + a_2)y(\vec{r}) = \left(\frac{k_1}{r} + k_2\right)f(r)\vec{r} \tag{2.19}$$

where k_1 and k_2 are some constants associated with the potential. Such equations cannot be easily solved in Cartesian coordinates as the radius of a sphere is defined as:

$$r^2 = x^2 + y^2 + z^2$$

and they depend on all three coordinates so they are solvable in spherical polar coordinates where every coordinate is an independent coordinate. Such equations cannot be easily solved in Cartesian coordinates as the radius of a sphere is defined as a coordinate independent of any angular coordinate. In general, an integral in Cartesian coordinates can be converted into spherical polar coordinates and cylindrical coordinates by using an appropriate choice of coordinates for ∇ as given in the last chapter:

$$\int_{-\infty}^{\infty} f(X)dX = \int_{-\infty}^{\infty} X(x)dx \int_{-\infty}^{\infty} Y(y)dy \int_{-\infty}^{\infty} Z(z)dz$$
$$= \int_{0}^{\infty} f(R)dR \int_{-1}^{+1} f(\cos\theta)d(\cos\theta) \int_{0}^{2\pi} f(\phi)d\phi$$
$$= \int_{0}^{\infty} f(\rho)\rho^2 d\rho \int_{0}^{\pi} f(\theta)d\theta \int_{0}^{\pi/2} f(\phi)\sin\phi d\phi$$

The equation of motion in Earth's gravitational field is relatively simple as we can always treat gravitational pull in the z-direction. This type of gravitational radial potential is expressed in a three-dimensional equation below as:

$$\frac{d^2 f}{dx^2} + \frac{d^2 f}{dy^2} + \frac{d^2 f}{dz^2} = \frac{f}{z}$$

If the length of the four-momentum is represented as P_{tot}^2, such that:

$$p_x^2 + p_y^2 + p_z^2 - \frac{E^2}{c^2} = P_{\text{tot}}^2$$

These differential equations are usually so specific to the physical system that the solutions of such equations will be discussed when they appear in real physical systems. The only specific case of an inhomogeneous equation is related to an infinite source and their solutions are given by Green's function as will be discussed later.

2.3 Lagrangian formalism

The applied force is known to navigate the motion of a system. The effect of force is measured from the applied potential. In moving fundamental particles, the effect of fundamental force is determined from the potential due to the presence of other particles in the system. Force applied due to the presence of another particle in the system is determined from the impact of the force due to another interacting object and creates potential for an incoming object. The total energy of the system is then called the Hamiltonian which includes all forms of energy including the kinetic and potential energy and is required to be conserved. However, dynamics is controlled by the energy that is available to perform an action. This available energy depends on the kinetic energy of the object in the presence of the applied potential. The total energy available to perform any work is determined by subtracting the potential energy V out of its kinetic energy T. This net energy L is classically defined as the Lagrangian of a system. This net energy is available to perform an action and can be identified as the energy of a system which can be used to perform a work (or instantaneous action) at any given instant of time.

Particles are identified by their intrinsic properties like mass and charge, and they acquire kinetic energy and momentum if affected by any external force. The dynamics of a system is determined by the applied force and the nature of the object and the Lagrangian gives the net effect of the interaction by including both the kinetic and potential energy. So, the Lagrangian formalism indicates the impact of interaction (potential). At the microscopic level, this interaction is related to fundamental forces which are associated by the intrinsic properties of particles. Therefore, the invariance of the Lagrangian under certain transformations is analyzed in the Lagrangian formalism using Hamilton's variational principle for dynamic equations of a system to determine the symmetries and define the related conservation rules. The Lagrangian of each interaction is expressed in terms of its potential and the invariance of the Lagrangian gives the symmetries of the Lagrangian, which is associated with its conservation rules. This Lagrangian formalism is derived from the constant action principle using the variational principle for many coordinate systems.

2.3.1 Invariance of the Lagrangian and conservation rules

The Lagrangian is defined as the difference between the kinetic and potential energy and determines the net energy of a system which determines the action of a system as a net response of a system. For this purpose, we can define the Lagrangian of a system in terms of kinetic and potential energy. For example, if we consider a hypothetical satellite of mass m revolving around Earth and ignore any other effect for the moment, the Lagrangian L is expressed in terms of the position vector $R \equiv R(r, \theta, \phi)$ in spherical coordinates and in the gravitational field. The Lagrangian as a function of spacial coordinates, spatial velocities and time is written as:

$$L \equiv L(r, \theta, \phi, \dot{r}, \dot{\theta}, \dot{\phi}; t) = \frac{1}{2}m\dot{R}^2 - mgR(r) \tag{2.20}$$

because the kinetic energy is a function of spatial velocities, whereas the potential energy depends on spacial coordinates. If the system is in dynamical equilibrium, the variation in the Langrangian is written as:

$$\delta L = \frac{\partial L}{\partial r}\delta r + \frac{\partial L}{\partial \theta}\delta\theta + \frac{\partial L}{\partial \phi}\delta\phi + \frac{\partial L}{\partial \dot{r}}\delta\dot{r} + \frac{\partial L}{\partial \dot{\theta}}\delta\dot{\theta} + \frac{\partial L}{\partial r\dot{\phi}}\delta\dot{\phi} + \frac{\partial L}{\partial t}\delta t \tag{2.21}$$

At this point, we can write the above equation in a compact form using generalized coordinates q that indicate any set of independent coordinates, in any coordinate system or a vector space:

$$\delta L = \frac{\partial L}{\partial q}\delta q + \frac{\partial L}{\partial \dot{q}}\delta\dot{q} + \frac{\partial L}{\partial t}\delta t \tag{2.22}$$

The Lagrangian can then give the information of interaction and can easily be used to understand the dynamics of a system. It is also worth mentioning that general information about the applied force and the nature of the force can be determined from the properties of the Lagrangian. In other words, the impact of a force or its potential on a physical system can tell the nature of the applied force including the symmetries and conservation rules associated with the applied potential.

2.3.2 Symmetries and conservation rules

The principle of least action tells us that the minimum energy required to perform an action is equal to the difference of energy between the initial and final state. Quantum mechanics of a single particle is in no way similar to the classical description of single-particle mechanics. However, we can use the principle of least action (or stationary-action principle) and the Lagrangian formalism of classical mechanics to solve problems of more complicated quantum mechanical systems. Before the discussion of this approach, we need to define a system by a set of n degrees of freedom called generalized coordinates q_j where $j = 1, 2, 3, \ldots, n$. All of these coordinates are mutually independent. We can define dynamics of such a system by a function of generalized coordinates q_j and generalized velocities $\dot{q}_j$ (the time derivative of q_j) as: $f(q_j, \dot{q}_j; t)$ and $\dot{q}_j = \frac{\partial q_j}{\partial t}$ The energies of a system can be defined in this form. Any action performed by a system is determined in terms of the available energy at a given time t. Just to understand the dynamics of a system, we define a quantity called Lagrangian L such that:

$$\delta L(q_j, \dot{q}_j; t) = \delta(T(\dot{q}_j) - V(q, \dot{q}_j)) = 0 \tag{2.23}$$

V can be a function of both q_j or $\dot{q}_j$ or even a constant. Net energy L can perform an action S in a given interval of time dt as $\delta S = Ldt$. The Lagrangian can also be the ability to perform an action at any given instant of time, whereas the total action is calculated from a Lagrangian as:

$$S = \int_0^t Ldt \tag{2.24}$$

Now the ability of a system to be able to perform an action at a given time t can be calculated using the principle of least action as:

$$\delta S = \delta \int_0^t Ldt = 0 \tag{2.25}$$

such that the principle of least action shows that the variation in Lagrangian δL at a given time t satisfies the relation:

$$\delta L = \delta \left[\frac{\partial L}{\partial q_j} + \frac{d}{dt} \frac{\partial L}{\partial \dot{q}_j} \right] = 0 \tag{2.26}$$

This condition is only satisfied if the term in parenthesis vanishes such that the minimum action remains constant:

$$\frac{\partial L}{\partial q_j} + \frac{d}{dt}\frac{\partial L}{\partial \dot{q}_j} = \text{constant} \tag{2.27}$$

Equation (2.27) is identified as the famous Euler–Lagrange equation which gives a set of equations corresponding to each generalized coordinate q_j as a separate conservation rule. These equations can then be solved as independent equations. This equation provides a powerful tool to study dynamics of a physical system in general including the low-energy classical limit to relativistic energy. It is also important to note if $V \neq V(q_j, \dot{q}_j)$ then equation (2.27) reads as:

$$\frac{d}{dt}\frac{\partial L}{\partial \dot{q}_j} = \text{constant}$$

then q_j is called a cyclic coordinate and it implies that:

$$\frac{\partial L}{\partial \dot{q}_j} = \text{constant}.$$

Physical coordinates associated with $\dot{q}_j$ represent a set of conserved coordinates. For example, if q_j represents the displacement in the x-direction, then the velocity in that direction or the associated momentum p_x is conserved. A linearly moving object with a constant potential exhibits spatial symmetry and leads to the conservation of linear momentum.

Since q_j's make a set of generalized coordinates, they are required to be orthogonal to be mutually independent of one another. This formalism can even be applied to quantum mechanical systems and various techniques can be applied to solve these equations. Basic concepts of quantum mechanics are discussed and applied to understand the Euler–Lagrange equations.

We deal with tiny systems and individual particle interactions in QED and the majority of such systems exhibit relativistic motions. Therefore, the interactions among particles are studied in four-dimensional space. These interactions take place at certain time in space. Dynamics of such systems is described by four-momentum (energy–momentum together) in four-dimensional coordinate space (space and time). Instead of talking about energy and moment conservation separately, we discuss the four-momenta conservation and the mass–energy transformation is incorporated following Einstein's theory of relativity. Einstein's equation of relativity, $E = mc^2$, defines a relationship between energy and mass. It further tells us that the energy of the incoming states is transformed into the momenta of outgoing particles and the transformation of momenta occurs via transformation matrices. Energy can either be converted into mass or vice versa. Properties of these transformation matrices help us to understand the nature of interactions. Contraction of tensors assures the existence of interactions through the assignment of covariant and contravariant vectors to incoming and outgoing particle states. Interaction is an action of force and Lagrangian formalism describes the mechanism of interaction. The Euler–Lagrange equations can be rewritten as:

$$\frac{\partial L}{\partial q_j} = -\frac{d}{dt}\frac{\partial L}{\partial \dot{q}_j} \tag{2.28}$$

which states that the variation in L with respect to the generalized coordinates q_j is equal and opposite to the rate of change of the derivative of the Lagrangian L with respect to $\dot{q}_j$. Since the kinetic energy T is just a function of generalized velocities $\dot{q}_j$ and $V \neq V(q_j)$ we find:

$$\frac{d}{dt}\frac{\partial L}{\partial \dot{q}_j} = 0$$

showing that the conservation of the Lagrangian is associated with the concept of a cyclic coordinate. Cyclic coordinates describe the symmetries of the Lagrangian under certain operation and it corresponds to a conserved quantity. For example, a translational invariance of the Lagrangian is written as:

$$L(x_1, m\dot{x}_1;\, t) = L(x_2, m\dot{x}_2;\, t)$$

and it states that the Lagrangian or ability to perform an action for a system remains unchanged even if a system is moved from point 1 to 2, giving:

$$\frac{d}{dt}\left(\frac{dL}{dp_1}\right) = \frac{d}{dt}\left(\frac{dL}{dp_2}\right)$$

which leads to the conservation of linear momentum $\frac{dp}{dt} = 0$. Linear momentum P corresponds to a cyclic coordinate in this case and it has to be conserved during translation to satisfy the above equation. We rephrase it as the invariance of the Lagrangian under translation or translation symmetry of the system which leads to the conservation of linear momentum. This simple example is generalized in terms of a relation between symmetries and conservation rules and plays a big role in understanding the various interactions.

2.3.3 Unitary symmetries

A group of symmetries is called unitary symmetries and is more relevant for fundamental interactions. Classically, conservation rules are associated with the equations of constraints such as momentum conservation which can only be employed if the linear translation is constrained to the objects with constant mass or the conservation of momentum is imposed to define elastic collisions. A detailed discussion of the representation of unitary groups and its symmetries is out of the scope of this book.

2.3.4 Lagrangian formalism and various interactions

In this section, we discuss the role of Lagrangian formalism in describing the nature of an interaction. Lagrangians of various interactions describe all the properties of the related interaction. Invariance of the Lagrangian under certain transformations describe symmetries of the interaction which is described by the Lagrangian. The corresponding Euler–Lagrange equation leads to rules related to the cyclic coordinates of the Lagrangian, defining conservation rules of the interaction. Every interaction is described by a particular Lagrangian and the symmetries and conservation rules of the Lagrangian define the nature of the relevant interaction.

Euler–Lagrange equations as equations of motion of a system are SDEs and are commonly-appearing differential equations in various fields of physics including quantum mechanics and electrodynamics. A linear differential equation is a one-dimensional polynomial equation defined as:

$$a_0 + a_1(x)y + a_2 y^{/} + a_2(x)y^{//} + \cdots + a_n(x)y^n = 0 \tag{2.29}$$

It is known that dynamics of most of the physical systems can be described as SDEs. Poisson's and Laplace's equations of electrodynamics describe the properties of interacting charges and their dynamics. Schrödinger's equation in quantum mechanics, on the other hand, describes the equation of motion of tiny objects at the quantum scale using wave–particle duality. So Schrödinger's equation along with Poisson's and Laplace's equations are equations of motion of an electromagnetically interacting quantum mechanical system. The dynamics of such a system can be understood from the solution of the relevant differential equations along a line, on a surface or in a three-dimensional space.

General solutions of SDEs describe the equation of motion of these objects using various interactions. We can look at a couple of generic equations in various coordinate systems. Starting with a one-dimensional general differential equation, the dynamic behavior of a function $f(x)$ can be understood by the following equation. We consider a general one-dimensional equation for k as a constant of motion:

$$\frac{d^2f}{dx^2} = k^2 f \tag{2.30}$$

this equation can be reduced to two FDEs by rewriting it as:

$$\left(\frac{df}{dx} - ik\right)\left(\frac{df}{dx} + ik\right)x = 0$$

giving two solutions:

$$f_1(x) = C_1 e^{ikx} = C_1[\cos(kx) + i\sin(kx)]$$

and

$$f_2(x) = C_2 e^{-ikx} = C_2[\cos(kx) - i\sin(kx)]$$

whereas a general solution can be written as:

$$f = A_1\cos(kx) + A_2\sin(kx).$$

This type of approach can even be used in two- or three-dimensional space as well. If the separation of variables can be managed to the level of rewriting 'f' as a combination of three independent functions:

$$f = X(x)Y(y)Z(z)$$

Thus, one obtains three independent equations which satisfy the individual wave equations in each coordinate for zero potential:

$$f = e^{\iota(k_x x + k_y y + k_z z)}$$

The same solution can become way more complicated in polar coordinates and we have to use special functions as a solution. These solutions even lead to the concept of quantization and describe atomic orbitals discussed later in the quantum mechanics chapter. Application of boundary conditions on these equations along with these solutions help to evaluate these unknown constant C's in quantum mechanics and associated conservation rules. This technique is further extended to understand the dynamics of various physical systems as well. Assuming that these function f's are analytical functions, they are considered to be single-valued differentiable analytical functions. They possess the same value inside and outside the boundary as a single solution and their derivatives are equated for at boundaries to find the correct solution of the equation.

2.3.5 Equations of motion in quantum field theory

The study of dynamics of particles moving at high energy involves relativity with quantum mechanics. We start with the Schrödinger equation again and express energy E in terms of the fourth coordinate of momentum, as discussed in chapter 1:

$$(x, y, z; ct) \rightarrow (p_x, py, p_z; E/c) \tag{2.31}$$

where c is the speed of light. The operator formalism of quantum mechanics is translated into four-dimensional space by representing the energy operator as a derivative of time. If time and energy are taken as imaginary coordinates, we call this natural coordinate space with real space and imaginary time Euclidean space. However, Minkowski, for convenience, rotated this coordinate system by an angle $\frac{3\pi}{2}$ clockwise to obtain a coordinate system with real time or energy and then the spatial coordinates (x, y, z) or corresponding three-momentum coordinates are imaginary components of the four-vector. This rotated space is called Wick's rotation and the new space with real time and imaginary space is called Minkowski space. It became a useful coordinate system as energy and time measurements for relativistic systems was more convenient than measurements of location and momentum simultaneously. The four-dimensional coordinate space $(\overrightarrow{x};ct)$ and the four-dimensional momentum space $(\overrightarrow{p};e/c)$ are conjugate spaces and the uncertainty principle relates $\overrightarrow{x}$ and $\overrightarrow{p}$ as $\Delta x_i \Delta p_i \sim \hbar$. Similarly, $\Delta E \Delta t \sim \hbar$ indicates set of conjugate variables. Basic concepts of quantum mechanics and relativity, in reference to QFT, will be discussed in later chapters.

2.4 Green's functions

Green's functions are special functions that appear as mathematical solutions to differential equations which have an infinite source. For the discussion of Green's functions, we need to understand the properties of the Dirac delta which can represent an infinite potential source. However, sometimes, more complicated situations may arise. Toroidal and poloidal motion are good examples of such unusual situations but they are not our focus right now. However, before going to Green's functions we first need to define a very important function, called the Dirac delta function. Green's functions give the ability to find a solution around an infinite potential source and the Green's function equation is written in terms of the delta function.

2.4.1 Dirac delta

We can start with the second-order derivation of the radial potential which is related to a function called the Dirac delta such that:

$$\nabla^2 \frac{1}{r} = -4\pi\delta(x, y, z) \tag{2.32}$$

The Dirac delta is an infinite function. The most well-known form of the one-dimensional form is written as:

$$\delta(x - a) = \frac{1}{2\pi}\int_{-\infty}^{\infty} e^{\iota kx} dk \tag{2.33}$$

This k is a conjugate variable to x. The Dirac function is equal to infinity at $x = 0$, and this infinity is related to the presence of a real source. Therefore, delta functions exhibit a specific nature and are very useful functions in quantum mechanics. For finite values of the limits of integration, this relation is also satisfied, giving:

$$\delta_n(x) = \frac{\sin(nx)}{\pi x} = \frac{1}{2\pi}\int_{-n}^{n} e^{\iota xq} dq \tag{2.34}$$

However, it has other representations as well and corresponds to infinite distributions. Before getting into properties the of delta function, we can look at a few other representations of the delta function which can be related to the above representation as well:

$$\delta(x) = \frac{1}{2}\frac{d^2}{dx^2}|x|$$

$$\delta(x) = \frac{1}{\pi^2}\int_{-\infty}^{\infty} \frac{dk}{k(k - x)}$$

$$\delta(x) = \frac{1}{\pi}\lim_{\varepsilon \to 0} \frac{\varepsilon}{x^2 + \varepsilon^2}$$

$$\delta(x) = \lim_{N \to \infty} \frac{\mathrm{Sin} N\pi}{x\pi}$$

$$\delta(x) = \lim_{\varepsilon \to 0} \frac{1}{\sqrt{2\pi}\,\varepsilon} e^{-\frac{x^2}{2\varepsilon^2}}$$

$$\delta(x^2 - a^2) = \frac{1}{2|a|}[\delta(x + a) + \delta(x - a)]$$

$$\delta(x) = \lim_{\varepsilon \to 0} \frac{1}{2\varepsilon}[\Theta(x + \varepsilon) - \Theta(x - \varepsilon)]$$

where $\Theta(x)$ is a usual step function. This delta function has very peculiar properties that can easily be checked:

$$\delta(-x) = \delta(x)$$
$$(x - a)^n \delta(x - a) = 0$$
$$\int_{-\infty}^{\infty} \delta(x - a) f(x) = f(a)$$
$$\int_{-\infty}^{\infty} \delta(\alpha x) = \int_{-\infty}^{\infty} \delta(u) \frac{du}{|\alpha|}$$
$$\int_{b}^{c} f(x)\delta'(x - a) = -f'(a)$$
$$\int_{-\infty}^{\infty} \delta(\eta - x)\delta(x - \zeta)dx = \delta(\zeta - \eta)$$

When we have a delta function for a higher-dimensional variable such as $\vec{R} = (x, y, z)$ and a constant vector $\vec{A} = (a_x, a_y, a_z)$, then a delta function attains the form:

$$\delta(R - A) \equiv \delta(x - a_x)\delta(y - a_y)\delta(z - a_z)$$

and can be generalized to n-dimensional vectors. For example, if R is a four-dimensional vector then we can express it in terms of four-vectors $r^\mu = (R, t)$ and the constant vector 4 has a constant time $a^\mu = (A, t_0)$, such that:

$$\delta(r^\mu - a^\mu) = \delta(R - A)\delta(t - t_0)$$

and so on.

2.4.2 Solution of differential equations using Green's functions

Electromagnetic interaction between two particles can be described as the equation of one particle in the presence of the potential of another particle. If these two particles undergo no changes and only the trajectory of motion is changed, we can treat it as a constant interaction or a finite radial potential where the minimum distance between two particles is maintained. However, when scattering processes take place, then they generate a nonhomogeneous equation as the energy and momentum conservation involve interconversion of energy into mass and vice versa.

A general form of the inhomogeneous SDE (in radial direction) can be written as:

$$\frac{d^2y(r)}{dr^2} + a_1\frac{dy(r)}{dr} + a_2\frac{1}{r}y(r) = f(r) \qquad (2.35)$$

where a_1 and a_2 are some unknown constants. The corresponding homogeneous equation is written as:

$$\frac{d^2y_c(r)}{dr^2} + a_1\frac{dy_c(r)}{dr} + a_2\frac{1}{r}y_c(r) = 0$$

where y_c is a complementary solution for the above equation. Since the radial potential $\simeq(1/r)$ goes to infinity at $r = 0$, it gives an inhomogeneous solution where the response of the system due to the existence of this radial potential can be found. The right-hand side of the inhomogeneous equation then reads $f(r) = \delta(r)$. We can therefore rewrite the above inhomogeneous equation as:

$$\frac{d^2G(r)}{dr^2} + a_1\frac{dG(r)}{dr} + a_2\frac{1}{r}G(r) = \delta(r)$$

and its solution $G(r)$ is called Green's function. In the case of a general source, the solution of the equation $u(r)$ can be written as:

$$u(r) = \int f(r')G(r', r)dr'$$

For a general source, the solution of the initial value problem, the convolution of $(G(r)^*f(r))$, gives a solution of the equation $Ly(r) + f(r) = 0$ in the presence of a potential $V(r)$. If we have $Ly(r) = f(x)$, then $y(x)$ is determined by the convolution of two functions $(G(x), f(x))$. A three-dimensional Green's function can be written as:

$$G(\overrightarrow{r} - \overrightarrow{r}') = \frac{1}{(2\pi)^3}\int \frac{e^{\iota kq(\overrightarrow{r} - \overrightarrow{r}')}}{k^2 - q^2}d^3q \tag{2.36}$$

$$= \frac{1}{(2\pi)^3}\int e^{\iota kq(\overrightarrow{r}' - \overrightarrow{r})}G(\overrightarrow{q})d^3q \tag{2.37}$$

The above equation satisfies:

$$(\nabla^2 + k^2)G(\overrightarrow{r} - \overrightarrow{r}') = \delta(\overrightarrow{r} - \overrightarrow{r}') \tag{2.38}$$

which is a well-known equation in quantum mechanics and will be discussed in chapter 6.

2.4.3 Retarded and advanced Green's functions

The solution of equation (2.36) has a double pole at $k^2 = q^2$. This integral can be solved using complex integration methods to solve this equation. The denominator of the right-hand side of equation (2.36) can then be written as:

$$\frac{1}{(k^2 - q^2)} = \frac{1}{(k - q)(k + q)} = \frac{1}{2k}\left(\frac{1}{k + q} + \frac{1}{k - q}\right)$$

which will help to change a double pole into two simple poles such that it has two simple poles at $k = q$ and the other one at $k = -q$. Complex variables can be used to solve the integral of equation (34) which has two pole points. These two poles can be removed using Cauchy–Riemann integration. We can then evaluate two forms of Green's function as:

$$G_{\pm}(\overrightarrow{r} - \overrightarrow{r}') = \frac{e^{\pm\iota k|\overrightarrow{r} - \overrightarrow{r}'|}}{4\pi|\overrightarrow{r} - \overrightarrow{r}'|} \tag{2.39}$$

The Green's function $G_+(\vec{r}-\vec{r}')$ corresponds to the outgoing wave, whereas $G_-(\vec{r}' - \vec{r})$ corresponds to the retarded Green's functions. Both of these functions converge onto the source point r'. Outgoing waves skip the interaction with the source, whereas the incoming wave is going to interact with the source and is the only wave that will have an effect from the source. These forms of Green's function are called the advanced Green's function $G_+(\vec{r}-\vec{r}')$ corresponding to the outgoing wave, and the retarded Green's function $G_-(\vec{r}' - \vec{r}')$ corresponding to the incoming wave. The incoming wave is going to interact with the source and its dynamics will be affected by the source.

In relativistic quantum mechanics, the four-dimensional formalism is used. In this case, the time-dependent part of the Green's function plays a big role in understanding the effect of interaction. In this case the physical interpretation of Green's functions is possible as the advanced Green's function gives a wave that has already passed by the source and the retarded Green's function is the wave that is about to interact, which correspond to outgoing and incoming waves in the interaction range, respectively. It is physically more important to find out the change in a wave after going through the interaction.

IOP Publishing

Conceptual Approach to Quantum Electrodynamics

Samina S Masood

Chapter 3

Computational tools

3.1 Introduction

Computation provides a very efficient tool for the precise calculation to find practical solutions for physical problems. The computer is an electronic device built on switching circuits developed by digital electronics. Current computers use Boolean algebra based on these switching circuits. In the language of computer scientists, it can be described in terms of bits and bytes. Bits are written in terms of 0 and 1, whereas a byte is defined as a unit made up of 8 bits incorporating all possible combinations of 2 bits and has a combination of eight digital circuits such as 10110011. Now bits are commonly used to describe the rate of data transfer, whereas bytes give the size of storage and memory of a computer device. The efficiency of computer devices and the storage capacity of computers has been increased significantly with the development of nanotechnology and its size has been reduced tremendously with the use of microchips and so on.

Microchips are included in all commonly used computer devices, such as smart phones, smart cars, digital projectors and small appliances. Now there is a next step in the development of computer technology which is using quantum mechanical phenomena of physics to develop quantum computing. This newer technology will increase the efficiency of computers almost exponentially. A quantum computer will increase all of the computational skill and the precision of calculation by tremendously reducing the required computational time and allowing huge data sets to be processed quickly. Understanding of quantum entanglement and other quantum phenomena helps in encryption and quantum teleportation as well.

Computers have developed graphing skill which led to the development of digital images and digital videos later on. These imaging techniques and video creation capabilities led to the development of digital media as a modern branch of media created by digital technology. On the other hand, the latest development of artificial intelligence (AI) is not just a technology, it is also a transformation of realities and facts into digital realities described using AI as a tool. AI provides machines which

doi:10.1088/978-0-7503-6054-8ch3

can understand the needs of users and then choose the most appropriate commands to work with. This new development is now being incorporated in computer programming at all different levels. However, the discussion of all these recent developments of computers using quantum computation and the development of AI are out of scope of this book. But they will be relevant for scientific computation ultimately.

3.1.1 Operating systems

Science is developed through experimentation, whereas mathematics and computational techniques provide effective tools to extract information from scientific data. Before getting into the computational tools, it is important to understand the basic operating system (OS) in the computational world. An operating system is a combination of hardware and software. Each operating system has its own hardware and software specifications. These operating systems are all built on the basis of system requirement and the relevant software which are particularly developed for the relevant operating system. The hardware part of an operating system is a subject of electronics and is out of scope of this book. However, all the operating systems have to be introduced to discuss the scope of scientific computing in relation to the operating system. We will just discuss those operating systems which are commonly used for scientific computing.

The most commonly known computers are personal computers (PCs). We call the associated OS Windows. Microsoft is the developer of windows. They are common and affordable. However, being so common, they are easily accessible for public use and can be hacked easily as well. The Android OS of cell phones and tablets is an offshoot of windows as well. A Windows-based OS is relatively more user-friendly and mainly based on graphic user interface (GUI)-based tools.

The second relatively less common well-known system is Macintosh (Mac). It has its own OS and is distinctly different from Windows. However, most of its applications are free and its working ability is stronger. It creates a set of file formats different than Windows. However, now Mac files can be transformed in a certain way in Windows and vice versa. iPhones, iPads and other Apple products use an offshoot of this OS. These two operating systems are run on two different machines. Mac machines are relatively difficult to hack. Their graphing card is especially very good and is a powerful system for visual applications. However, it is a little more expensive and its file transformation to Windows may not be a direct transformation. You may need to modify files properly to make them readable during transformation and before saving into a newer format in new OS. The scope of both systems is slightly different and they are distinguishable clearly.

There is a third OS known as UNIX with totally different scope. It has totally command-based OS and almost comes from the original version of the computer when basic computer users started typing in text file and began perfoming computation. This OS has more technical importance and handles more serious group computing than commercial uses. UNIX is the OS for powerful computing and helps in parallel computing or in the construction of supercomputers. It has the ability to handle large programs which can be run for days and a huge set of data can be analyzed using data mining and other much more advanced special methods. UNIX-based programs provide the most effective computational tools for researchers, as especially scientists and engineers have to use it. However, these days, several UNIX-based programs can be commanded by windows or Mac systems as well. Unix has a totally unique set of file formats. The only transferable files are mainly Portable Document Files (PDFs).

Linux was developed to make UNIX easily understandable for users. However, graphic files have several common formats. Linux is a GUI-based version of UNIX and is becoming popular among users. Some of the UNIX-based programs can be run on Windows or Mac, after transforming them into a Linux version. Linux generates simple Unix files and creates small files, which considerably saves computer space. We will introduce some commonly used computational tools and the associated environment or software in this chapter without getting into details of OS.

Since most of the commonly used computers are Windows-based, all the software and languages are integrated with Windows or Mac. Most of the computer languages are command based. These commands are very well understood by UNIX or Linux easily. But they are translated to any OS using relevant drivers. Usually all the packages have different drivers or installers for different OSs. Just for installation of software a relevant installer for every system has to be installed using the correct executing program. All the languages are translated using a compiler which can translate a high-level code into a low-level language to be understood and executed by any OS. Android and Windows have compatible files, and similarly, iPhone and MAC have compatibility. Now programs are developed to translate and read files from other OSs. There is now a lot of success in this direction.

3.1.2 Computational tools

Integration of experimental study with mathematical and computational techniques is behind all the scientific discoveries and inventions. However, the proper choice of required approach is extremely helpful. In fact, all three approaches provide a certain level of accuracy about scientific realities which leads to new explorations and improvement in existing information as well. Mathematics emphasizes the generalization of results which helps to make rules and understand basic principles, whereas computational techniques provide accuracy and precision. In short, experiments, mathematical calculations and computational analysis all play a crucial role in understanding scientific principles and their applications, which lead to new discoveries and inventions. It is a commonly known fact that the proper choice of appropriate tools at correct time provides great help in creating a high-quality project with maximum utility. Similarly, confirmation of any phenomenon depends on the confirmation of results using various approaches and reproducing the same results more than once. Actual findings have to be reproducible and independent of space and time. These findings are accepted after reproducing them using various approaches and techniques.

Analytical solutions are always preferred because they can give exact results which can easily be applied anywhere with the proper definition of variables. Analytical calculations are extremely important if the relevant equations are solvable exactly. However, mathematics has its own limitations. Several well-known equations describing dynamics of a system such as quadratic equations and regular differential equations may not be solvable analytically and alternate methods are developed to obtain the best possible solutions. However, every physical problem cannot be expressed by a set of equations that can be solved analytically to find an exact solution of the problem. These complicated equations usually have a large number of parameters. An exact solution of these equations is only possible if the number of parameters matches the number of equations. Otherwise, all equations are evaluated for various sets of the unknown parameters to generate mathematical data and that data can be analyzed to find the most probable solution. These results can then be compared with the experimental/observational results or an experiment is designed to test the theoretical results. In case of disagreement between experiment and theory, an experimental design is modified to be able to find a better agreement between theory and experiments. In this way, going back and forth between calculations and experiments is done to find an agreement between theory and experiments.

A graph of computer-generated data can then be interpreted and extrapolated to physical phenomena outside the experimental range as well. It helps to design further experiments and develop hypotheses for physics beyond the range of existing technology. So a computational approach not only understands the current form of a system, but also its past and future behavior. A combination of mathematical equations and the computational data analysis then describe the basic mechanism behind the creation of a system and its dynamical behavior. These equations cannot be analytically solved for each system and sometimes the calculations become very long and complicated due to the presence of several terms and a large number of unknown parameters.

Computational methods can perform large calculations in less time, more accurately, and in a more foolproof way. We do not have to worry too much to recheck the calculation, which saves a lot of time. There are several commonly used examples for computational methods, which include solutions of complicated differential equations or multiplication of a large set of matrices, we can still solve it by substitution methods. If there are more unknown parameters than the number of equations, we can still obtain a relatively easily solvable equation and choose one or a minimum number of unknown parameters. Then as a next step, we can solve these equations numerically. Sometimes some of the unknown parameters can be fixed with the experimentally determined range to provide a cutoff for unknown parameters.

Mathematics itself has a natural limitation due to its complexity and the required human concentration and calculational skill. Moreover, a variation in magnitude of working scale of various systems and complexity of the behavior makes analytical calculations more challenging. The computer has provided tremendous help to develop parallel methods of calculations. Computers can solve equations precisely from nanoscale to the cosmic scale. Calculations of single equations at various scales may sometimes require us to choose a different set of approximations to solve equations.

Computers can be very helpful when calculations become too challenging and data analysis is required to understand the behavior of physical systems. This is especially true when numerical values are needed, as computation becomes more helpful. Numerical data can be generated with computers and dynamical behavior is understood by plotting this data. It does not take away the need for analytical calculations. Numerical data is easily generated by insertion of numerical values and can even be done by hand but not so efficiently. On the other hand, computers help to analyze the existing data and generate working models for the dynamical behavior of complex systems just based on the actual experimental or observational data. This is how computers are used for building models of existing or under-construction machines. Mathematical tools are required to investigate their properties and working ability. These mathematical/computational models with their relevance for natural objects provide the biggest learning tool and can later be used in various ways in the development of new technology.

Mathematical equations can cover a large range of sizes and measure times from nanoseconds to millions and billions of years, whereas computed parameters may have their own ranges which may not match with the mathematical range. Nature has a great variation in sizes and shapes with a long range of variation in time. However, much more specialized mathematical or computational tools may be needed to reach the detailed study of extremely large and infinitely small systems. Technology requirements may vary for differing measurements. Regular units are more relevant for our daily life, whereas the scope of a machine is not limited to standard units. For detailed structure analysis, extremely minute observations are needed, whereas to look at the cosmos, we have extremely large separations in space and we talk about huge distances and the time scale is much longer than our own lifetime. Scales range from nanometers (10^{-9} m) to billions (10^{9}) of meters or even beyond this range.

When the detailed calculations seem to be too cumbersome to manage easily and further details become out of scope of mathematics, a parallel computational branch is developed to understand continuous behavior by using numerical approximations. Using numerical methods, we can solve much more complicated problems. Especially long expressions are solved in a relatively convenient method using numerical approximation. For example, multiplication of several gamma matrices can be done by incorporating the properties of gamma matrices, multi-dimensional integrals and complicated calculations with several steps. Some of the simple existing packages such as Mathematica, Mapple and MATLAB are very commonly used for mathematical calculations. They are used to solve complicated integrals as well. These packages are successfully used by engineers or scientists. Almost every graduate student of mathematical sciences has to use these computational methods.

Numerical analysis and computational techniques are also used to get help in analytical calculations including integration, differentiation, matrix multiplication and other complex mathematical calculations. Computational methods are also used in model building based on the experimental data or observational results. These methods are not limited to a single academic discipline. One computational method including data analysis or data mining techniques could prove equally good in entirely different disciplines like economics, bio-science, astronomy or even sociology. However, they are needed to solve complicated problems and definitely increase the scope of application of physics to complicated physical systems studied with totally different approaches.

Numerical methods and other computational techniques need a very long discussion as we can use alternate methods to obtain required results. Available software packages can help to some extent and can be modified to an extent but they have their own limitations. On the other hand, original codes can be written using a relevant computer language. All these methods are so specific that describing each and every package or including the discussion of any suitable language needs practical experience. We will just discuss the names of computer languages and only a few software packages.

3.1.3 Programming languages

Programming languages are the languages which are used to communicate with computers. The choice of the language depends on the fact of what it is needed for. These languages are translated in terms of the combination of bits and bytes to give commands to computers to operate the way a programmer wants. These languages work in different OSs such as Windows and Mac, which are commonly used these days. Other smart devices such as iphones and Androids for phones and cars and other devices have their own OSs and have specially developed Apps to communicate with computers. There are a large number of programming languages including Python, Pascal, Fortran, Basic, C++ and machine language that are used to set up a computer. All of the software packages are developed based on these languages.

3.2 Mathematical typing

Mathematics is a language of communication which has its own vocabulary, consisting of mathematical symbols and Latin and Greek indices to represent parameters, whereas its sentences are written in the form of equations, indicating the rules and principles which describe the behavior of a system. Therefore, the clarity of mathematical symbols and equations is extremely important to study a physical system. In physics mathematics becomes even more important when we do not have direct access to a system. It provides a kind of tool to virtually access a system and mathematically describes the behavior controlled by certain principles. Therefore, the correct appearance and clear representation of equations is very important in mathematics.

Mathematical symbols and equations are not easy to type within English or any other language alone because they include particular mathematical symbols in addition to Greek indices, whereas moving equations around or alignment of more complicated (summations, integration, subscripts, superscripts, etc) equations with text as well as among its own terms becomes challenging and it becomes difficult to represent them in a readable way. At this point, LaTeX provides particular tools needed to represent equations in a well-organized way with symbols written clearly. Microsoft Office also includes symbols and a template of equations to be able to write simple equations in Microsoft Word and other applications now. MS Word also has an add-on for math typing called MathType which provides a lot of symbols, but still does not cover each and every symbol. MS Word provides convenient visual tools to format regular typing and it is very easy to learn. However, when it comes to large and complicated equations, LaTeX is needed to overcome the limitations and challenges of Word. Another inconvenience is that Word does not move symbols and text around together and editing of a file can become very time-consuming.

Mathematical typing packages are usually developed in a Linux environment using LaTeX. Now even Windows-based LaTeX packages are available and math-typing packages can be included to Microsoft as well. Mathematical computation is still considered to be more convenient in Linux or UNIX environments. Linux and mathematical programming can be done in Mac as well. However, now Windows-based LaTeX packages are available and online resources can be used in Windows for programming. Moreover, Microsoft Word and LaTeX files can now be copied and pasted into each other. Tables and figures can be saved as usable files in LaTeX. Mathematical symbols are part of Microsoft Office as well. Typing of simple mathematical equations is now very convenient even in Microsoft, especially Word and PowerPoint. However, complicated mathematical expressions are not easy to write in Microsoft Office. Now Windows-based LaTeX packages are available online as a free download and are easy to learn with a little practice. We introduce basic parts of LaTeX programming briefly, but it can be easily learned using online resources. LaTeX provides many mathematical symbols, which helps to write well-organized equations.

LaTeX files use regular keyboard and mathematical symbols and Greek letters are typed as words in English usually. Even online sources are available. In a way, LaTeX is a type of package which has its own syntax like a language. It is added as software that is associated with a repository which can convert LaTeX commands into mathematical symbols or organize them in the required format. A repository is a central location in which data is stored and then organized. This is how the mathematical commands are converted from LaTeX to PDF format that can be read properly.

LaTeX files start with a class file with the first statement describing the document class. These files include all formatting information and connect to the required packages from the repository. A LaTeX file includes a 'usepackage' command in the beginning to use various packages from a repository. LaTeX files are then executed to give a log file and may produce PostScript files or PDF files. Using appropriate packages in LaTeX, the mathematical equations and symbols can be easily typed and organized following proper methods. It includes special self-directed tutorials for typing mathematics in LaTeX. A list of commands to type symbols or organize equations is available online along with others. On the other hand, a few online resources are available for using the LaTeX commands in an appropriate way.

LaTeX is originally a UNIX-based program, which is totally a command-based language. It uses commands to type math and compile the mathematical commands into equations. It has been accepted as the most convenient method to type long complicated equations. Therefore, LaTeX packages are developed for other operating systems such as Windows and Macintosh. A GUI-based package for Windows called Scientific Work Place or Word Scientific was created, but it could not get enough popularity due to its cost and certain limitations. Almost all packages are mainly designed for high-energy physics and have built-in Feynman diagrams and have the ability to reduce gamma matrices.

As mentioned earlier, LaTeX generates a log file that can be converted into a PostScipt file and/or the more commonly used PDF format. It can also be opened as a text file in any operating system. They can be converted into any acceptable file format there. It is also worth mentioning that LaTeX gives an option to create presentations as well. However, they are created in PDF format. Regular English can be easily copied from MS Word and pasted into LaTeX or the other way around. Text can now be transported by copying from one file type and pasting into another file type from even a different operating system.

3.2.1 Typing of equations

Mathematical equations and symbols can be typed and formatted in the math mode only. Mathematical symbols are usually defined in LaTeX in a particular form. They are case sensitive. Lowercase Greek symbols start with lowercase English letters and uppercase Greek symbols start with uppercase English letters. In a way, LaTeX is a language of mathematics in itself and expresses mathematics in a special way. Subscript, superscript, summation, equality sign and other mathematical symbols are written in a particular way in math mode and executed in PDF in the form of actual mathematical symbols very nicely. LaTeX is almost the only way to accommodate all mathematical symbols, nicely and properly aligning them with text and other symbols. Once mathematical equations are typed, their formatting can be easily modified by using appropriate commands for writing symbols and equations. This way we can copy the entire equation and paste it again and even modify them according to need. Text from the LaTeX file can be copied and pasted from or into Microsoft Office files or any other relevant file format as well.

In the math mode of LaTeX, mathematical equations are typed and arranged in an easily readable format. Matrices and determinants are typed in a convenient way. Even mathematical symbols can be included in matrices. Mathematical symbols and operations are also possible to type inside the matrices. Matrix equations can be typed and fitted in a proper way. Numbering of equations can be controlled very well by gathering several equations under one command and numbering it individually. A group of equations may carry just one number or each equation in a group can be numbered individually or left un-numbered. There are methods to control the orientation of equations. They can be written in the next line, along with the text and then control their orientation and alignment in a group of equations, their location in the center of left or right. Numbering can also be employed using various popular methods.

Matrices typing is a little challenging in several typing programs and fitting them properly in the text is an even bigger challenge. LaTeX allows matrices of any size in a proper way, using special command. However, they fit very well in equations in a properly lined way, as it appears in most of the modern mathematical books. For a complicated long equation, the mathematical terms can be managed properly and arranged in a proper order with LaTeX only. It is true that LaTeX learning may take a little time which may be very quick because even with the basic knowledge and downloading free LaTeX programs from the web, it is very easy to type anything using the internet sources which give a correct set of commands, packages needed to use those commands, and even their appearance in PDF format or PostScript files.

3.2.2 Figures and tables

Figures and tables are inserted in LaTeX easily. LaTeX generates PostScript files and figure files can be created or managed in PostScript format as well. These PostScript files can be fitted very nicely in the document. However, LaTeX allows figures and tables in several commonly known figure formats such as JPEG, PNG, TIFF, PS, and PDF. These figures can be labeled and a caption can be included during the figure insertion. PostScript files of figures can be converted in any size without compromising their pixel density. Unlike Word documents, labels always stay aligned with the figures. Tables can be added in a similar way with appropriate referencing. These figures and tables can also be resized.

LaTeX works in a Linux environment and it can import figures and tables from Linux-based programs even in PostScript format. It becomes very convenient to pick data, figures and tables from mathematical packages. An additional benefit of LaTeX is that we can directly type tables in LaTeX in a chosen format and label them properly. The labels and numbering stay there properly. Tables and figures can be imported from any other program by saving them in well-known figure formats. Excel data can be copied and saved as tables in PDF files and added as PDF file-like figures as well. In addition to all that, numerical results and plots can directly be inserted from mathematical programming.

3.2.3 Including a bibliography

Citations in LaTeX are typed in a separate file and numbered using special commands. Bibliography files can even be typed separately and added to any LaTeX file. Then the numbering of references can be manually included in the text as required. This usual process is the same as is adopted in Word. However, a very convenient way to use a bibliography in LaTeX is that each reference can be assigned an individual label. Numbering of references in bibliography files becomes foolproof and extremely convenient as LaTeX takes care of numbering automatically from its label and cannot be incorrect. There is no need to do individual numbering of references. It might appear to be a little more complicated but this is a very convenient method because later on these labels can be used everywhere without worrying about forgetting the numbering of references or their format or appearance in the text.

LaTeX is very helpful in organizing references and has an ability to convert their appearance according to any publishers' requirements just using the provided class files by publishers. This numbering and referring style is applicable to equations, tables and figures as well. All of the equations can be automatically numbered and labeled in a way that their referencing can be handled through labels as well. Equation numbering occurs automatically as they appear in the text. This is one of the very convenient features of LaTeX.

3.3 Scientific programming

Mathematical programming makes use of numerical methods, related algorithms, and specific codes for the problems under consideration. It incorporates the related algorithms based on the required principles and symmetries of the theory relevant for the physical systems. The scientific code under discussion is based on numerical methods needed to solve the mathematical part of the problem. Meanwhile, the other type of scientific programming involves data analysis and model building as well. This part of scientific computation is as important as mathematical calculations.

Scientific study and development is based on theoretical study, experimental observation and scientific data collected by various observations or experiments. Computational methods are developed to analyze all the collected data and analyze it thoroughly to understand the behavior of physical systems. This scientific data analysis leads to the understanding of mechanisms involved in the production of data. Choosing the correct computer language for a particular purpose is important, as some can be more useful than others.

If we just concentrate on two basic goals for scientific study, mathematical computation and data analysis are very different from each other. Mathematical computation needs distinct tools to analyze a particular type of data. Various computational packages are developed for different purposes based on main languages such as Fortran and C. Alternatively, data analysis involves statistical analysis and every package has its own scope which can make it better or worse based on the application. In fact, a correct and timely choice of an appropriate program or efficient language gives quicker and better results.

3.3.1 Scientific languages

There are various computer languages that can be used to write a computer code for mathematical computing. A very well-known language is Fortran (that was originally developed for formula translation). Fortran has its own syntax. A large number of physics labs and complicated computational research groups use Fortran as it has some very useful features which makes it a common language in the scientific community.

There are several other languages, developed afterwards, that can be used to create codes for mathematical computation. C++ and Python are the two important examples. Python is a relatively simple and modern language that is easy to understand. The syntax of various languages overlap but they use different compilers and they work under different logic. The choice of languages depends on the purpose of calculation. Big labs or large theory groups using mathematics and physics still prefer Fortran or C++. They are the most popular scientific computing languages, compared with some of the modern languages such as Pascal and Python and even the machine language. AI makes use of various languages to develop codes for particular tasks, and now there are methods to go into deep study as well.

Improved computer technology and programming techniques are regularly updating the computational techniques. However, some of the special tools related to a field of research have been around for a long time and continue going through upgrading to be able to accommodate the needs of recent developments in the field. Most of the mathematical packages used to use Fortran or C++. They are commonly used in the form of newer versions. Python is relatively more user-friendly language and it is more commonly adopted for coding. However, based on the requirements of the field, usually multiple languages are learned and used according to the need. However, there is a part of syntax which is adopted by every scientific language. Therefore, knowledge of one language makes the learning of others much easier due to the overlap of syntax and logic in every language. However, the difference of accuracy in calculational approach may change the preference to use one package over the other.

3.3.2 Software packages

In the current era of technology, big scientific developments are related to the group work performed in big international labs and large collaborations and interdisciplinary expertise are required to discover and invent new technology. Therefore, well-developed, communication becomes extremely important. For this purpose workstations are used as a common working environment for a working group where software, data and any other information can be shared among the group from distance as well. A much more developed form of workstation with a huge computational capacity is called a supercomputer and it can be shared by a huge group of people.

Much of the academically used software is available as free downloads as its developers get credit for the development. Almost every field has specialized tools that are learned while starting working in the field. A few well-known codes and packages are available for scientific study. Some software programs are commonly used in certain fields of study like the Cactus framework which provides resources for different fields of research regarding many-body problems. Its standard toolkit like the Einstein toolkit provides tools for the calculation of complex problems of numerical relativity and fluid dynamics. Cactus provides the ability for parallel computing, data distribution, and checkpointing. Additionally, Gaussian is a popular chemical physics program and is used to study rates of chemical reactions, making it a great tool to study molecular behavior. It is based on density functional theory and helps solve problems with a supercomputer or on a workstation as well.

Similarly, Mathematica, MATLAB and Maple are commonly available packages used for relatively simple scientific calculations. Some other particular packages are also available for special types of work. For example, FeynCalc is used for the calculation of Feynman diagrams or the reduction of gamma matrices, and QuantumATK is used to study molecular properties and nanostructures. Amber, NAMD, and Visual Molecular Dynamics are programs for the study of molecular dynamics and most of them are GUI-based, while others are command-based only. Computer-aided designing (CAD) software programs are available for engineers and are developed for various engineering fields and modified according to their needs.

There are several other programs based on special codes written for a particular type of calculation in various fields of research. Computational tools are more commonly used in theory and tremendously increase the scope of calculation and accuracy of results. We obviously cannot list all the existing tools for every field of study because it is a very long list. Moreover, software development is an ongoing process which involves upgrading and an increase in efficiency of existing packages in newer versions and improvement in the scope of computation. Newer software programs with new features are always introduced for more detailed scientific investigations or replacement of existing software.

Data analysis plays a crucial role in the study of complicated systems and is required to develop natural sciences, human sciences and social sciences. Computer scientists and engineers develop required machines to be able to analyze statistical data, and analysis methods were developed and learned and applied as needed. Similarly, by developing relevant statistical techniques, medical, biological, environmental, agricultural and astronomical data can be analyzed. Astrophysics, condensed matter physics, plasma physics and many other branches of applied sciences infer results from data analysis. Existing data provides a starting point for theoretical investigation as well. In other words, the approaches of theory, observation, and experimentation all play their roles, interchangeably, to understand physical behavior and develop science and technology. Computation can then invariably be used in every stage of development and common languages provide a means of communication among computers as well as among scientists, especially if common framework or various features of the same software can serve the purpose at different stages of investigation. This is usually a motivation behind an upgrade or different version of the same software.

Most of these packages provide their own list of commands. So there are several modified languages to develop particular tasks in given languages. The Printer Command Language (PCL) was developed by Hewlett-Packard specifically to allow computers and printers to communicate with each other. The PCL files consist of commands. HP printers parse and decode those commands. The PCL format also supports HPGL plotter files (PLT). These languages are specifically used to communicate with certain devices for the same purpose and not for computational purposes. It is also important to mention that these days self-learning tutorials and workable examples of several packages are available to learn software. These tutorials and examples are presented by software developers and work with the package very well. They usually come for free with the packages and some of them may be developed for more specific needs and can be purchased. Online training and troubleshooting tools may also be available.

3.3.3 Scope of various software

We live in a huge ocean of knowledge and quick learning ability is required to stay up to date. It may be possible to become deeply involved in the field of research without quick learning, but it works very well to develop widespread knowledge for a sound background. Most software is designed to provide results in various forms such as numerical results. We can ask the computer to give that result which we need and ignore any other unwanted information regardless of its importance. Therefore, the same software can be used for various purposes. Even the change of parameters may make it usable by several disciplines.

Most commonly used output formats are numerical data in the form of tables, graphs, figures and videos. Simple programming can be considered as an improved version of scientific calculators. Computation becomes important as it involves a series of different steps and solves several mathematical equations simultaneously to generate theoretical data in the form of a table or create a graph of data for the given range of variables. Its cosmetics include color, thickness, shape of data points and so on. We ignore cosmetics for now and focus on the purpose of data generation. This tabulated data can be put in the form of a graph, which may be possible to plot manually as well. But the accuracy level may not be the same. However, computers graphs can be organized in several forms, as needed.

Computer graphics is a multipurpose tool. It highlights the special features of data in a graphical form and it can easily show special features of the functional behavior for a particular range. We can look into the highlighted feature in more detail by exploring numerical data or re-evaluate for smaller variation of input parameters to get more detail. So graphical analysis provides a quick way to reach into detailed analysis of the behavior of a system. Now we can create graphs very quickly with a small change of data to understand the behavior of data in more detail to uncover the hidden features.

Computers can create figures instead of graphs as well, which can be considered as an extension of graphing. Figures can be based on equations instead of their particular values. We can change data quickly and multiple figures can be created quickly. We know that the quick change of figures can create motion pictures and turn them into videos. This is how we can create digital videos in addition to optical videos as well. When we get the ability of creating video format from equations, this is where design enters computing and gives birth to graphic design as well. We will not be able to discuss these topics here.

Development of visual output ability of a computer has started an extremely new feature of powerful computing. Now the dynamics of physical systems can be visualized from a set of equations in the form of videos. This visual simulation of data is a very powerful way to learn dynamics of a physical system. The computational method creates computer models for complex systems using a set of equations. Simulation is a very powerful tool of computing and is more related to understanding the dynamics and visualizing it.

3.4 Mathematical computing

Mathematical computing in itself is a branch of computer science that helps in developing programs for computing. Due to the complexity of mathematics in detailed study of physics problems, computational methods are developed to solve using computers. Initially to solve several equations simultaneously, even knowing mathematical techniques, computer programs save time and increase the scope of calculation. Just based on the mathematical logic, an algorithm is developed using a set of equations and computationally performs all the required mathematical steps to reach the result relatively quickly and solve it on the first attempt, once you test the accuracy of the code to your satisfaction. This code may be designed to incorporate various approximations as needed. The choice of variables and their ranges are always related to the physical system being studied and choose a set of allowed approximations without compromising the required information to an unacceptable level.

Mathematical logic, numerical methods and an algorithm work together to solve a mathematical equation on computer. A computational approach looks different but it works very well to solve definite integrals and equations in a physical range. Another built-in requirement is the existence of an analytical solution. An equation or a set of equations can only be solved for a definite problem using a finite set of parameters with the analytical range only. Computers cannot yet handle several mathematical issues, and infinities is one of them. Therefore, the solution of indefinite integrals is not possible as they may have infinities. Moreover, any singularities of a theory have to be removed manually. Some of the other rules such as the correct order of multiplication of matrices or fixing the order of various binary operations or other special rules for solving a given set of equations for a particular purpose require special commands to be provided to the computer through its algorithm to perform the required process properly.

3.4.1 Numerical methods

Numerical methods provide a set of computational techniques which are developed to solve mathematical equations by computer for physical systems. Numerical methods are based on the simple form of calculational processes which are used to solve these equations in a simpler way, adoptable by computer. It may adopt a longer but simpler method. Analytical methods give a complete solution with or without physical approximations. Numerical methods are efficient when correct analytical solutions are not possible. Step-by-step evaluation of a series of numerical values of a function corresponding to the limited required range of the given parameters, helps to understand the behavior of a function based on the computer-generated data within the given range. It does not give a complete solution of equations but it can tell the behavior obtained for a given range of variables and can be used and even compared with experimental or observational results to validate the original equation for a given range. Regardless of the fact that a theoretical equation may not be correct to describe the behavior of a system, in general, it could nevertheless still be adopted for the given region of interest. For the range of values for parameters in the region of physical interest, an approximate solution can be obtained. Therefore, the numerical methods evaluate equations for a particular range and a numerical solution can be found for the given range of values only and not a complete solution which can be used for any system.

Numerical methods develop some mathematical steps that can be translated to computers to solve mathematical equations. This method is very helpful to develop effective tools to understand the system in a workable way. Unknown parameters are usually estimated by solving a comparable number of equations and a computer can even help to solve a large number of equations in less time, which can be solved by hand but take much less time and do not need to be rechecked multiple times after making sure that it is working properly. Since several problems (in physics and engineering) cannot be solved analytically due to their complexity, practical solutions can be found using such techniques.

Numerical computing uses simple analytical mathematics based on arithmetic operations using any suitable programming languages such as Python, Fortran, C, C++, or available mathematical packages such as MATLAB, Mathematica or Maple. All these packages have their built-in logic and help to solve a set of equations quickly. However, these packages may come with certain limitations and the logic behind the operations may not be fully known for these packages; this may sometimes lead to a result which is more approximate or oversimplified than desired. Therefore, scientists have to develop a more customized code instead of using packages for more detailed investigations. Sometimes other kind of problems can create limitations such as using well-known coordinate systems because of their unusual shape and size or an unusual complex behavior which can be described properly using well-known laws. Such issues can conveniently be resolved by choosing valid approximation which leads to a calculable solution.

Numerical calculations can be done by evaluating equations for given values of variables and plotting various functions (quantities) simultaneously. The behavior of unknown parameters can then be extracted from their plots and even compared among themselves. Simultaneous solutions of equations can be obtained from numerical solution of a set of equations for relevant variables. This way, a computer helps to solve problems for a particular system even if a general solution cannot be found and figure out the behavior within the particular situation.

3.4.2 Algorithm development

Programming languages are used to develop computer algorithms for the purpose of developing a computational technique to solve a particular research problem incorporating the underlying assumptions and relevant approximations.

An algorithm is based on a complete sequence of required equations obtained for a specific problem using all the relevant conditions and approximations. It is used as a step-by-step evaluation of a set of relevant equations to get a particular solution in the form of quantitative results

An algorithm describes a step-by-step method based on basic principles to understand the behavior of a system by solving a mathematical problem incorporating basic principles and related symmetries and conservation rules. The constraint equations provide restrictions or segregate between allowed and forbidden approaches to solve problems. When needed, constraint equations provide extra equations to evaluate more unknowns from the given set of equations, solving equations simultaneously as the conservation rules provide additional relationships between various variables and help to solve a problem for more unknown parameters. An algorithm has the ability to incorporate all valid approximations when applied to a particular system. A good algorithm should not use any approximation and leave it to the user while a code is written for a given system.

A simple form of an algorithm may be graphically represented as a flowchart and describe the logic behind the computational process. A flowchart is useful for simple calculations, whereas an algorithm describes the logic behind complicated processes as well. In other words, a flowchart can be generalized to an algorithm and can included more details of the involved process. Therefore, an algorithm provides logic behind a computational process and is applied to solve physical problems.

In an algorithm, one can develop a generic program which can be altered according to physical systems and correct descriptions of required parameters. It has more room to change parameters going from one to another physical problem. It is not too specific to be useful for only one system. It is generally applicable to a certain type of problem with substitution of correct variables. The logic of an algorithm is based on basic principles. Then using an algorithm, a code can be written for particular problems. These codes are written in one particular computer language and are based on the logic which is suitable for particular tasks.

3.4.3 Code development

A practical computer code is then developed according to the proposed algorithm. A general algorithm is a set of commands which describe all the steps involved in computation in the same language. It may leave the option to leave the initial values of parameters open and get the results for a particular problem where all the parameters are evaluated corresponding to the independent variable. Moreover, an algorithm is converted to a code in an understandable computer language. In case needed, sometimes the syntax of various mathematical languages may be close enough to let a short code convert from one language to another and use it for different applications with minimal effort.

An algorithm has a basic logic needed to write a computer program. It refers to the logic behind the computational technique. A code, on the other hand, is a set of commands that tells a computer exactly what to do and how to solve a given problem. A computer code can specifically may be more appropriate for a specified range and can be modified for a different range. Numerical algorithms can then be developed to visually describe the processes in the form of graphs or diagrams. A set of diagrams can be converted into videos as well. A salient feature of a code is that it includes all commands for a computer to give the results in an understandable human language and printable format.

Results obtained by a code can be altered for various values and ranges of the parameters and graphs and videos can be obtained as required. In addition to just printable numerical results or even mathematical forms, it is also to be mentioned that sometimes the content of the algorithm and code can be interchangeable. Numerical methods are used to develop algorithms to solve a particular set of equations. A set of equations in an algorithm can be solved with minimum possible approximation, if not exactly. Some of the algorithms are already based on some approximations to avoid getting into the situation where equations are not solvable any more. Therefore, the choice of the applied approximation has to be done carefully.

Once an algorithm is developed, various computer languages can be used to write a code to communicate with the computer. The entire code has to be written in the same language. The execution of mathematical operation directed by the code is used to generate theoretical data involving the results corresponding to the provided parameters as controlled variables. A code may have the ability to give the result in the form of an equation, if chosen. Computer-generated numerical data appear in the form of tables or graphs as well.

Depending on the language and the available equations, numerical results can be obtained in the form of a table including the provided parameters and the corresponding results in the form of a table. A detailed format of the table can be chosen giving appropriate commands. This data can be plotted for a required range of variables or can be obtained for various ranges. We can choose to plot graphs for various ranges of parameters in chosen colors. These graphs can be plotted in various forms More elaborated forms of results could be figures, movies or even simulations. However, these results can only be generated based on the scope of the code, computer efficiency, appropriate algorithm and proper language.

3.4.4 Data analysis

We had been discussing theoretical computation and undoubtedly computers are extensively used in theoretical study. Theoretical physics is now heavily based on computation. Special branches of physics such as known numerical relativity, high-energy physics, fluid mechanics, plasma physics, material science and many other theoretical physics branches are fully tied up with computational study. The same is true for other science disciplines and engineering designing. Development of science is fully indebted to computational tools, these days. However, the computer has provided a totally new approach in research which is called data analysis.

Data analysis provides a very effective tool to understand the behavior of a physical system from experimental measurement. Data analysis is an efficient tool to deduce results from the experimental data. It is almost the only way to understand physical systems that are either too complex to unfold the detailed behavior like a living system or totally out of reach, such as heavenly bodies or nuclear matter. In other words, where experimentation is not possible and it is too risky. For these out of reach complex systems, experimental data provides a working tool to understand the behavior closely. Based on the experimental or observational data, various models can be constructed to describe the dynamics of a system. This approach is called data analysis.

Statistical analysis and small tools are provided to make data analysis as effective as possible. For this process a few underlying data processing methods are applied such as removal of noise, optimization of data, curve fitting and study of the logarithmic behavior, etc. Data analysis is an extremely powerful technique in medical science where it can help to learn about the behavior of the human body under life-threatening situations, connection with the environment, astronomical model building to learn about the cosmos and getting deep into the nature and behavior of nuclear matter.

Various data analysis techniques are used for complex systems. Like other software, various data analysis packages are developed to study special classes of systems. Special packages are available to use a particular strategy, which is more relevant for an under-consideration system to obtain correct results. Astronomical Data Analysis, an interface description language or interface definition language (IDL), is used to communicate with the computer. This is a generic term for a language that lets a program or object written in one language communicate with another program written in an unknown language. IDLs are usually used to describe data types and interfaces in a language-independent way, for example, between those written in C++ and those written in Java.

Statistics is the discipline that concerns the collection, organization, analysis, interpretation, and presentation of data. In applying statistical methods to scientific, industrial, or social data, it is conventional to begin with a statistically acceptable approach and a working statistical model is applied for such studies. Regression is a set of statistical modeling that is applied to statistical processes for estimating the relationships between a dependent variable and one or more independent variables. Quantitative analysis is a complex topic, and it deals with several steps like medians, modes, correlation and regression.

In statistics, special packages are available for data analysis. A few popular ones are Excel (MS Office), Origin and Tableau. Some of the computer languages have their own software in their own languages. R is an open source program for data analysis. Python has the ability to analyze data. Jupyter Notebook (Linux), Cactus, and even C++ have their own packages for data analysis. Now data analysis itself is merging as a new computational field as data science. It gives an approach to analyze existing data collected from various sources and summarize their trends and main characteristics, often using statistical graphics and other data visualization methods. Almost all common regularly used software now have the ability to analyze data and get different forms.

It is worth mentioning that the development of computational techniques is so fast these days that it is very hard to describe them in small sections or in a chapter.

We can just recommend interested students to look for specialized literature for detailed study and practical experience is even more important. Moreover, all the modern big research facilities develop their own computer codes and share their own software to protect their intellectual property and secure their own findings. These extremely specific packages are only developed for big labs and specifically designed research facilities. They can choose one of the available languages and the start to develop a package, just for convenience. These packages continue for decades and usually remain in the same language. However, knowing one language properly and having enough experience moving around various languages allows one to choose appropriate software and receive comparable results.

Analyzed data can then be put in a model through graphing the data and performing error analysis and statistical approach. Once a simulation provides a technique to visual understand the visual behavior of available data on an observable scale. Rescaling may reduce cosmic scales to a computer screen or expand the micro-scale to a computer screen and convert billions of years or nanoseconds to minutes. This way we can visualize extremely small and short-term processes and cosmic processes of billions of light years away objects to visible scale.

Simulation can be considered as one of the output forms of data analysis. It is the most convenient form of the output to virtually visualize the output similar to what you can get in real time. However, it takes much more computer time as compared to the simplest form of the output as a single numerical value. We can get more data through programming for various sets of input parameters and get results tabulated in a printable form.

A graphic card helps to create various forms graphs of data and plot data into two-dimensional forms as contour graphs as well. On the other hand, media cards can be used to generate videos based on figures obtained from data analysis. These graphic cards and video cards employ special programs themselves. A larger set of data helps to develop better pictures with more pixel density and better-quality videos can be made such as high-definition videos. Visual output is a very effective source of information and helps to conceptually visualize the dynamics of a system.

Simulation helps a lot when we try to create scientific theoretical models for practically unreachable physical systems as small as subatomic or sub-nuclear systems or genetic modification. Examples are large systems such as the weather-related spacial models which can be created using a set of fluid equations or more complicated systems such as the merging of stars, supernova explosions, or the Big Bang model. It simply helps to uncover the secrets of inaccessible physical systems. Simulating a larger data set helps to explore the physical systems in much more detail. Human beings are now getting more into data science and data mining. Quantum computation and AI are newly emerging computer-based resources to handle much more complicated computation for this purpose. All of these techniques are more efficient than all the existing technology and going through further development to be able to store even more data and analyze larger data packages quickly to reach new results and applications. More data statistically gives more accurate results.

On the other hand, to protect your own intellectual property, security of personal information and safety measures become more important and challenging due to its vulnerability. Therefore, data encryption becomes equally important and is continuously developed to fulfill the needs of large data. So a common language for a facility is protected by developing their own programs for big facilities and the efficiency in computation within shared facilities is obtained by developing a powerful shared common computational system utilizing devices such as supercomputers. Smaller organizations can develop a relatively simpler system by developing a common workstation that can have the ability to run a code as a shared cyber-physical system (CPS) from different PCs at the same time. However, development of password, image scanning, teleportation and encryption have become a part of computational research. Now even quantum computing is being developed to improve computational scope.

In the end we will conclude by saying that computer technology is advancing through two ways. One is the development of computational methods including numerical analysis, algorithm development and code writing. Newer approaches such as data science, AI and quantum computing are also advanced versions of computational techniques. On the other hand, hardware development is joining hands with software development to advance computer technology. Newer graphics cards, quantum computers, and media developers are going through development swiftly. Computer technology advanced from semiconductors to liquid crystals, and now nanotechnology is helping further the development from improved equipment and newer codes.

Part II

Relativity and electromagnetism

IOP Publishing

Conceptual Approach to Quantum Electrodynamics

Samina S Masood

Chapter 4

Electromagnetism

4.1 Introduction

Matter is found as an electrically neutral macroscopic bulk material around us. It consists of various compounds as a collection of molecules, which can hardly be found in their pure state. These molecules are made up of atoms which are considered to be the fundamental building blocks of matter and represent the chemical properties of elements. Atoms are electrically neutral and participate in each and every chemical reaction to make all the matter around us. Molecules, on the other hand, are the smallest units of compounds which can exist independently and carry all the properties of compounds. They are made up of atoms but physical properties of atoms are not recognized in molecules.

Atoms and molecules are electrically neutral but they are composed of charged particles called electrons and protons, and electromagnetic interaction plays a fundamental role in all chemical processes which helps to facilitate all material modifications. Some of the complex molecules can be decomposed into ions or radicals under certain conditions with nonzero charges. Atoms and molecules are spinless neutral electrical particles. All the chemical processes are governed by a specific form of electromagnetic interaction and describe the chemical bonding and molecules formation. Ionic motion, inter-atomic and inter-molecular forces are described in the form of Van der Waals force and several other chemical phenomena can be described in terms of the motion of charged particles. Basic electromagnetic theory is the interaction between charges which are pretty well-known at the macroscopic level and is a long-range force just like gravity.

Electromagnetism is the only successful fundamental theory which works perfectly at the microscopic and macroscopic levels simultaneously and takes part in all chemical reactions. The molecular form of all matter is made up of atoms and ions or radicals. It acquires different forms to facilitate various kinds of processes including organic and inorganic chemical processes, and even the biochemical processes. This is the only interaction in which fundamental principles of

doi:10.1088/978-0-7503-6054-8ch4

electromagnetism are indebted to the properties of charges and the related fields. The electric field is intrinsically associated with charges and the magnetic field is dynamically generated from moving charges. Some of the fundamental principles of electromagnetism are introduced in undergraduate textbooks in one-dimensional form and it is relatively easy to understand basic concepts adequately. Solution of three-dimensional equations is much more complicated, especially in spherical polar coordinates as spherical symmetry works better with the central force. Cylindrical shapes are more complicated due to different symmetry, whereas irregular distributions of charges are even more complicated to understand the overall effect. Dependence of electromagnetic force on charge distribution makes the shapes of electric components much more relevant to study the macroscopic form of electromagnetism.

4.1.1 Detection of charge and electromagnetism

In 1791, Luigi Galvani discovered that a charge can travel through muscles because when frogs touched the metallic surfaces, their muscles are contracted, indicating that the electric charge travels through muscles and causes the contraction of muscles. Not only that, the attraction between oppositely charged surfaces and repulsion between the same charges was noticed without knowing the reason behind it. Alessandro Volta then started the study of properties of charge through a series of experiments using different metals like iron, lead, tin, zinc, silver and graphite, etc, and study the flow of charge through the common metals.

History of electromagnetism starts from 600 BC when Greeks (probably Thales of Miletus, a mathematician) found that a yellow translucent mineral called amber develops the ability to attract tiny and lightweight objects such as feathers and hair, when rubbed with fur. The name 'Elektron' is actually amber in Greek. However, the concept of two charges and the interaction between charges was not known at that point or the information was not probably documented for another 23 centuries. Since this phenomenon involves rubbing, it is also called triboelectric effect. This shows the relevance of electromagnetic interaction in life and material properties.

In the 18th century, Benjamin Franklin established the conventional use of two types of charges, negative and positive charges. It was discovered that opposite electrical charges attract each other, whereas similar charges repel each other. The interaction of charges plays a key role in creating the macroscopic form of matter. Charles Coulomb, afterwards, started to quantify charge in a standard unit called the coulomb. Today we know that the fundamental unit of electric charge is the electron, and one coulomb is made up of 6.24×10^{18} electrons. Conventionally, electron charge is considered negative and proton has an equal and opposite positive charge $e = 1.6 \times 10^{-19}$ coulombs. The smallest independently existing fundamental particle is the electron which has a negative charge, set as a unit charge.

Electromagnetism describes the combined effects of electricity and magnetism which were found to be related to each other and were unified via Maxwell's equations. This is a unique force which is equally important at all the observable scales from the microscopic to the macroscopic world. Moreover, it plays a crucial role in the existence of life and development of technology at large scale and maintains the structure of the fundamental building blocks of matter: atoms and molecules.

Electricity is a commonly observed phenomenon which gives a description of several well-known natural observations. It explains the theory behind lightning (produced when two oppositely charged regions happen to interact to destroy their charge to light), St. Elmo's fire (a space weather phenomenon of creation of luminous plasma due to coronal discharge), electric discharge due to a group of torpedo fish (type of cartilaginous fish, commonly known as electric ray) the amber effect (amber when rubbed with a feather, can attract lightweight objects), etc. All of these observations were not understood until the discovery of electric charge and the interaction between them was discovered. Equation (4.1) describes a relation of magnetic field generated by a constant electric current which is produced by moving charges. It relates the magnetic field to the magnitude, direction, length, and proximity of the electric current. Moreover, though the electromagnetic force is spherical in nature, the distribution of electric and magnetic fields will depend on the geometry of charged objects or the geometry of the electromagnetic setup.

Later on, it was found that an isolated electric charge could propagate through certain metals which were identified as conductors later. Charge may be stored in a capacitor. The first charge-storage device was designed as an early form of a parallel plate capacitor and called Leyden jar, named after its designer (figure 4.1). It was basically a glass jar, containing a few sheets of metal foils. A pierced cork was used

Figure 4.1. Leyden Jar the first charge-storage device.

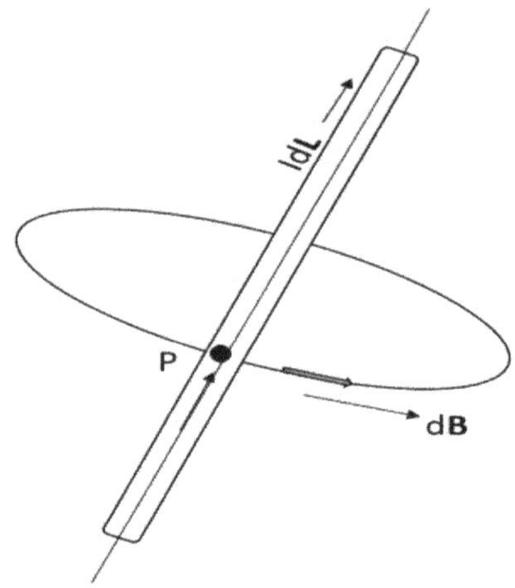

Figure 4.2. dB is the magnetic field realized at point P from the current element IdL.

to close the top of the jar. Figure 4.2 represents the creation of magnetic field from current The exposed end of the wire is brought in contact with a friction device to charge the jar. This charge will be collected and stored between the inside and outside of the glass jar under high potentials. Invention of the charge-storage device was a very important discovery in electromagnetism and played a key role in the development of technology, especially information technology.

Energy can be converted from one form into another form and when human beings learned how to make use of this conversion, new devices were made. Batteries were invented to convert chemical energy into electrical energy. They can be considered as a small chemical reactor that can store chemical energy in between two electrodes. These chemicals produce highly energetic electrons that can flow as current through external devices and current is defined as the rate of flow of charge. The discovery of static charge and the knowledge of the flow of current led to the invention of the incandescent lamp, which opened a gateway to technology.

This process did not stop here. Batteries provided an external source of energy that could be used to run different devices. When batteries were available to initiate a process in a machine, we were able to construct several machines by converting electrical energy into mechanical energy. Further investigation of properties of charge and the flow of current led to the discovery of the basic principles of electromagnetism and Maxwell's equations then introduced the phenomenon of electromagnetic waves. These electromagnetic waves are used to define visible matter.

4.1.2 Basics of electromagnetic theory

Classical electromagnetism deals with the properties of electric charge and its ability to do work, which is associated with electric field E. Later on, with the discovery of electromagnetic waves, a relationship between electric field E and magnetic field B was discovered. Electric fields are a vector and can either be attractive or repulsive. A positive charge exerts a pulling force and a negative charge has the ability to exert pushing force but the net force depends on the polarity of both charges. Same charges repel each other due to competing forces and opposite charges attract each other as they support each other. Properties of static charges, net effect of charge due to their geometric arrangement and the overall effect of charges enclosed in a geometrical arrangement by a hypothetical surface is another interesting topic of electromagnetism.

Basic principles of electromagnetism were discovered from the detailed study of properties of static charge and the flow of charge as current. Basic principles were discovered relating the electric and magnetic field. Properties of static charges were related to electrostatics, whereas the magnetostatics deals with the properties of magnets created by certain arrangements of magnetic dipoles. Magnetic dipoles are a combination of two magnetic poles defined as south pole and north pole. Magnetic field lines coming out of the north pole and going into the south pole represent the magnetic force of attraction. The density of this magnetic flux B is called the magnetic field, in common language. The magnetic field lines between two opposite poles move in a circle indicating attraction, whereas the field lines between the same poles never overlap and result in repulsion. It was later found that magnets are produced due to the alignment of charges and are always arranged in the form of dipoles. North and south poles are generated together so the poles are always broken into dipoles.

Opposite electric charges and opposite magnetic poles are attracted to each other and similar charges and similar poles repel each other based on the strength of charges and poles and the separation between them. The similarity of electric charges and magnetic poles (charges) indicate a similarity between electricity and magnetism. The only obvious difference is that the electric charges are separable and one kind of charge can be found flowing as it is just as negatively charged electrons can flow in the form of current from positive to negative charge or higher to lower potential. However, magnetic poles are not separable. Magnets are always found as a dipole. This difference is related to the fact that charge is the intrinsic property of matter and cannot be removed from particles, whereas magnetic poles are not intrinsically associated with matter. Magnetic field is created by moving charges, instead, and can be switched off by stopping the motion of charge. However, some naturally existing materials may show magnetism due to the material structure.

The electric and magnetic forces are proportional to the strength of the corresponding charges (electric charges or magnetic charges) and inversely proportional to the square of the distance between them. The inverse square law of force between two charges and two poles is similar to the inverse square law of force between masses, named as gravity. Electric force, magnetic force and gravity are all spherical in nature and vary with the square of the distance between two relevant objects. Magnitude of gravity and electromagnetic force are proportional to the product of two masses and two charges, respectively. The constant of proportionality of these forces is different in both cases and determines the relative strength of forces.

However, these forces can be differentiated by their degrees of freedom. Gravity deals with masses and it has only one degree of freedom and produces an attractive force only. Charges and magnetic poles participate in electromagnetic interaction with individual degrees of freedom and can be attractive or repulsive in nature. Opposite charges attract each other and same charges repel each other. Electric fields are polarized whereas gravity is not polarized at all. Gravity acts from lighter to heavier mass as light objects are always attracted to heavy objects and tend to reduce the distance between masses. However, both of these forces are central forces and cause orbital formation at entirely different scales. Planetary orbitals are produced in space due to a competition between gravity and angular momentum, whereas the atomic orbitals are created due to a competition between electromagnetic force with the intrinsic angular momentum at subatomic scale.

Central forces are symmetric around the center of the sphere. One of the charges or the mass of the above-mentioned forces should be situated at the center. So the central forces discuss the motion of an object around another object exhibiting circular motion. The electric and gravitational forces are both central forces, which are uniform on the surface of a sphere as the magnitude goes with the square of the distance and not the direction. Both of these forces are spherically symmetric in nature and are responsible for orbital motion under favorable conditions. This spherical nature makes spherical polar coordinates as the most suitable coordinate system to properly understand these forces. With electromagnetism we study the overall properties of net charges which may not necessarily be the unit charges. Electromagnetism is more like a macroscopic version of electrodynamics and does not dig deep into the individual behavior of charges where the quantum mechanical approach becomes more relevant.

Gravity is always attractive but electromagnetic fields and potentials have two degrees (positive or negative) of freedom each. It deals with static charges and their properties. Dynamics of gravity work with basic conservation rules of physics and Newton's laws of motion. Kepler's laws of planetary orbitals also involve gravity. Electrodynamics is developed as a formalism to study the movement of charges with respect to one another in three-dimensional space and relate it to the external coordinate system. This technique develops a set of basic rules to study electrodynamics that is usually described as fundamental principles of electrodynamics and can relate electric and magnetic fields with respect to other charges which describe the relationship between them and their dependence on the rate of flow of charge or current.

Current is defined as the rate of flow of charge. The electromagnetic theory of radiation indicates that the moving charges have a time-varying electric field associated with them. Moreover, the moving charge may create a magnetic field perpendicular to the direction of motion. The magnetic field generated due to the current distribution involves a vector product between the direction of the flow of current along the current-carrying conductor and the distance from the current to the field point. The magnetic field of a current element $\vec{dB}$ can be expressed as:

$$\vec{dB} = \frac{\mu_0 I\vec{dL} \times \vec{l_r}}{4\pi r^2} \tag{4.1}$$

$\overrightarrow{dL}$ is the infinitesimal length of current-carrying conductor and $\overrightarrow{l_r}$ indicates the distance to the field point. The above relation between the magnetic field and its source of magnetic field is called Biot–Savart's law of electromagnetism. This law is expressed in terms of the distance from the current. The direction of the magnetic field is determined using the right-hand rule which states that if the fingers of the right hand are rolled around the thumb along the magnetic field, the thumb indicates the direction of the current.

Basic principles of electromagnetic theory are developed using measurable parameters of the theory and their relationship. Most of the features of the theory are studied and discussed at the macroscopic level in the lab environment and are related to the collective behavior of charge or the current flowing through certain materials and the impact of electromagnetic forces on the formation of electromagnetically-interacting systems. Classical electrodynamics is a slightly more complicated theory of an electromagnetically-interacting fluid but can still be considered parallel to fluid mechanics because the basic difference of nature is the main difference between them. Fluid mechanics considers mass of fluid molecules and gravitational force plays a pivotal role in the flow of liquid. Electromagnetic interaction is a much stronger force and surpasses gravity for light particles, whereas electrically neutral heavy particles are still under the dominant influence of gravity. The result of the competition between gravity and electrodynamics is what is a deciding feature to choose a better approach comparing fluid mechanics or electrodynamics.

Several rules are related to the collective behavior of charge. These well-known rules that we call fundamental principles are either found during empirical investigation or derived theoretically and then tested experimentally. Some of the fundamental laws of electromagnetism such as Gauss's law, Faraday's law of electromagnetic inductions, and Ampère's circuital law were initially discovered as behavior of charge and related the magnetic field with electric field. The relationship of fields with current was also mainly the motion of charge and the generation of electric and magnetic fields. Some of the other important relations like this were derived out of the fundamental principles as well. However, these fundamental principles led to the development of Maxwell's equations to understand the electromagnetic waves. These laws of electrodynamics are discussed in detail in the next section.

Electromagnetism is the only theory which has equal relevance for single-particle dynamics and appears as a collective theory of large macroscopic objects. Current is essentially defined as the rate of flow of electrons but it can be measured when a stream of electrons flows through a conductor, which could be a microscopic or a macroscopic object, and the the principles of electromagnetism are translated to macroscopic systems in terms of current and voltage.

4.1.3 Electromagnetic waves

In the presence of moving charges, electric and magnetic fields change with time and time-dependent fields are found to satisfy the wave equations:

$$\frac{\partial^2 E}{\partial x^2} - \frac{1}{c^2}\frac{\partial^2 E}{\partial t^2} = 0$$

and

$$\frac{\partial^2 B}{\partial x^2} - \frac{1}{c^2}\frac{\partial^2 B}{\partial t^2} = 0$$

and the solution of the above equations gives the form of electromagnetic waves as

$$E_x = E_m \sin(kx - \omega t)$$

and

$$B_x = B_m \sin(kx - \omega t)$$

where ω is the oscillating angular frequency of the electromagnetic wave given as $\omega = 2\pi f$. F is the frequency of waves such that $\frac{Em}{Bm} = c$ for c the speed of light which is similar to $x = v\ t$ where v is the velocity in the direction of x. The energy associated with the electromagnetic waves is indicated by a Poynting vector $\vec{S}$ given as:

$$\vec{S} = \frac{1}{\mu_0}\vec{E} \times \vec{B}$$

The energy transport with electromagnetic waves is a vector perpendicular to both fields and its magnitude can be given as:

$$S = \frac{1}{\mu_0}EB = \frac{1}{c\mu_0}E_m^2 \sin^2(kx - \omega t) \tag{4.2}$$

where we can use the wave forms of the electromagnetic field in terms of sinusoidal waves and the average of sine of the angle gives the net amount of energy as:

$$S = \frac{1}{2c\mu_0}E_m^2.$$

4.2 Flow of charge and electronics

Conservative forces do no work along a closed path and have to obey the relevant conservation rules. Mathematically, this means that any conservative force can be expressed as a gradient of a potential as $\vec{F} = \vec{\nabla}\ V$. Just as the gravitational potential describes the motion under gravity, mass always moves from higher to lower potential. Similarly, charge moves from high potential to low potential. However, the work done by the electric potential is zero. The study of electromagnetism is related to a stream of electrons as current and the difference of voltage (defined as per unit charge) and is defined in units of volts in SI. Voltage is a quantity that is related to the energy associated with a single charge and is directly related to the electron charge. The dynamics of current flow can only be understood in terms of current and voltage. Since current flows through materials and is used more in macroscopic systems, the relation between the current and voltage is integrated with the properties of materials and a whole new field of study has emerged as electronics which controls the technology and the industrial applications of electromagnetism. Fundamental laws of electronics control the flow of current in an electric circuit and several electric components have been developed to convert electrical energy into mechanical work or other circuits dealing with the close path of current that follow the basic conservation rules.

The rate of flow of charge defines the electric current I, which is used to transfer electric energy from one place to another place. This flow of current is facilitated by the difference of electric potential V defined as electric energy per unit charge. However, when the current flows through a medium, the properties of the medium affect the movement of electrons. Therefore, the rate flow of current significantly depends on the conducting properties of the medium. These materials are generally categorized as superconductors, conductors, semiconductors and insulators. Every material can be put in one of these four groups with its characteristic value of resistance. Ideally speaking, superconductors have zero resistance and the insulator has infinite resistance. Each solid and liquid material can have a measurable resistance R, which tells us how much current is allowed to pass through for a given potential difference.

Electronics uses electromagnetism to understand the flow of current for the applied voltage depending on the structure and properties of various materials. Commonly found resistors are usually Ohmic resistors which are directly related to the properties of conducting materials such as metals and electrolytes. However, the current and voltage are always proportional to each other and the ratio between the current and voltage remains constant for a given medium. This principle is called Ohm's law which is the basic principle of flow of current. If voltage V is measured in volts, and current I is measured in amperes then the ratio between the voltage and current is called resistance and is expressed in units of ohms (Ω), in standard SI units. Proportionality between current and voltage is called Ohm's law and can then be expressed as:

$$V = IR \tag{4.3}$$

An understanding of the dynamics of charge plays a pivotal role in the development of applied physics and technology and it is studied in physical electronics. It is based on the understanding of the flow of current due to a given difference of potential. It was discovered that the flow of charge induces a magnetic field which is associated with a perpendicular force called electromotive force (EMF) and is created by voltage supplies or electrochemical batteries. It can be expressed in terms of the magnitude of the perpendicular distance from the current-carrying wire or the path of the flow of current.

Efficiency of electronic components and their design is based on these basic principles. However, engineers use various combinations of circuits and several additional factors to choose the right material, size and shape to make it practically beneficial and cost-effective. A detailed calculation of EMF involves the basic principles of electromagnetism and will be discussed in detail wherever it will be more relevant. Principles of electronics developed another branch in electronics named circuit electronics, which overlaps with electric engineering and starting from developing small electronic gadgets and devices. Designing of the electronic appliances and complicated equipment is mainly done by engineers who have specialization in certain areas. An efficient design may be created considering material properties, electronics, market trends and appropriate use of available tools along with the creative skill.

The flow of current follow certain laws and provide a strong basis to develop circuit electronics and modern technology. Kirchhoff's rules are based on the principle of conservation of current in a closed loop and the conservation of voltage across nodes. They are called Kirchhoff's laws and are the basic principles of closed circuits. These laws tell that the loop currents are conserved to conserve the total charge. The flow of current towards a junction is equal to the flow of current outside the junction and the law is also called junction rule which can be written as the vanishing sum of incoming and outgoing currents. The Kirchhoff's junction rule states that the sum of all currents entering a junction is equal to the sum of all the outgoing currents:

$$\Sigma_i I_i = 0 \tag{4.4}$$

On the other hand, the conservation of electric energy demands that the total energy per unit charge or the sum of the potential difference around a loop remains unchanged or the change in potential around a closed loop is zero and is also known as the loop rule which can be written as:

$$\Sigma_k V_k = 0 \tag{4.5}$$

These two rules are basic principles of circuit electronics and they are applied to transfer electric energy into the required form. However, the electronic components such as resistors, capacitors, transistors, power supplies or electronic meters are designed using the concepts of electromagnetic materials. The same basic principles of electromagnetism are used to make power supplies, batteries, oscilloscopes, etc, and describe the mechanism of working of the basic components. The principles of using electromagnetism are studied to define the working of electronic components.

Development of modern technology is largely indebted to the flow of current through a combination of simple and complicated circuits which involve a variety of electronic components. Electronics is one of the basic applications of electromagnetism and plays a crucial role in the development of various forms of technology. Physical electronics deals with the mechanism behind the working of electronic components and the basic rules followed by these circuits, which help engineers to design electronic devices using circuit electronics. Conversion of electrical energy into mechanical energy is used in making mechanical tools and is very helpful in household appliances and small tools, whereas the conversion of electrical energy into other forms, which makes various kinds of tools and equipment, is a large part of technology. Digital circuits are also designed using circuit electronics, which is associated with the development of computers and information technology. Electronics itself is a vast topic and its detailed discussion is out of the scope of this book. However, the knowledge of physical electronics and electrodynamics has a big overlap. In this book we will skip the electronics part and move to electromagnetism now.

4.2.1 Electromagnetism and electronics

A static charge has an electric field $\vec{E}$ which can be defined as the ability of a charge to apply a force on a unit charge. Another associated field with a charge is created when a charge moves through a conductor. The associated field with a moving charge is defined as a displacement vector field $\vec{D}$, such that $\vec{E}$ and $\vec{D}$ are related to each other through the electric permittivity ε and $\vec{D} = \varepsilon\vec{E}$. The electric permittivity of a medium $\varepsilon = \varepsilon_0^*\varepsilon_r$, where ε_0 is the permittivity of free space and ε_r is the dielectric constant of a medium. On the other hand, a moving charge creates a magnetic field of strength $\vec{H}$ and the magnetic flux density $\vec{B}$, perpendicular to the direction of the moving charge. $\vec{B}$ and $\vec{H}$ are related through the magnetic permeability μ such that $\mu\vec{H} = \vec{B}$. The magnetic permeability of free space is given as μ_0.

Before getting into the discussion of laws of electromagnetism, we have to define a few more quantities such as the electric permittivity ε and the magnetic permeability μ. Electric permittivity is the property of a material to allow the electric field lines to pass through it. In other words, it measures the characteristic of a material and measures the opposition offered by a material against the flow of current. On the other hand, magnetic permeability is related to an imaginary porous material which can be described as an imaginary permeable surface which allows the magnetic field lines to pass through it. Magnetic permeability can then be related to the number density of those imaginary pores which allow the number of lines of magnetic force passing per unit area. For calculation purposes the electric permittivity in free space is defined as ε_0 and the magnetic permeability in free space as μ_0.

Electric permittivity and magnetic permeability are related to the properties of a medium. It is also important to notice that there are four fields altogether: electric field $\vec{E}$, electric displacement field $\vec{D}$, magnetic field $\vec{H}$ and magnetic flux density $\vec{B}$. Flux is defined as the number of lines of force passing through a unit area per unit time and $\vec{B}$ corresponds to the magnetic flux density. The electric field $\vec{E}$ is related to electric flux ϕ as $\vec{E} = -\vec{\nabla}\phi$.

4.2.2 Current–voltage relationship

The discovery of Maxwell's equations describes the role of moving charges in creating the electromagnetic energy in terms of electromagnetic waves. Non-relativistic motion of charge can transfer energy to materials to produce radiation and that gives a beginning to the modern technology which helps to convert electrical energy into mechanical energy. Ampère's circuital law with the conservation of electrical energy helped to develop the laws of electronics which are described as the conservation of nodal current and loop voltage in Kirchhoff's rules.

There are four fundamental laws of electrodynamics which make up Maxwell's equations and express electromagnetic waves in terms of electric and magnetic fields. Four basic principles of electrodynamics are briefly discussed here to develop Maxwell's equations. These laws include Gauss's law of electric and magnetic fields and then Faraday's law and Ampère's law. All of the other equations of electrodynamics can be derived from basic principles of electromagnetism and are found through experiments.

4.3 Electromagnetic theory

Electrostatics deals with static charges and Coulomb force is the force between static charges. There is an electric potential and an electric field associated with static charges. Moving charges can produce magnetic field and there are four basic principles of electrodynamics which are represented by Maxwell's equations. These laws include Gauss's law of electric and magnetic fields and then Faraday's law and Ampère's law. All of the other equations of electrodynamics can be derived from these four basic principles of electromagnetism and are found through experiments.

Electromagnetism is a combined theory of electricity and magnetism and is defined in terms of electric charges, magnetic dipoles, electric and magnetic fields and electromagnetic forces, including Coulomb force and Lorentz force. Coulomb force is the force exerted by a charge, intrinsically, whereas, Lorentz force is the forces associated with the moving charges and helps to define the magnetic field. Faraday's law and Ampère's law establish the relationship between the variation of electric and magnetic fields. These principles of electromagnetism provide a foundation to build electromagnetism which can further be developed into classical electrodynamics. The classical electrodynamics along with relativity and quantum mechanics gives a description of quantum field theory which makes it a successful quantum theory and it works simultaneously at large scales as a long-range theory such as gravity. It will be discussed in detail, later in this chapter.

Faraday's law and Ampère's law allow us to discuss electricity and magnetism as a manifestation of the same phenomenon. Moving electric charges inside a wire create a magnetic field around it. A moving magnet can also produce electric field. Electromagnetism then describes the properties of charges and deals with the associated electric and magnetic fields. It is a perfectly testable theory at the macroscopic level and usually deals with charges at the macroscopic level.

The fundamental interaction known as electromagnetism competes with gravity and is not only used to describe the atomic and molecular binding, but also plays a crucial role in the formation of large material structures and multi-particle complex systems. Electromagnetism alone describes the collective behavior of matter along with gravity and identified with different labels due to its role in the behavior of matter in all different situations. Properties of matter in various phases including solids, liquids and gases are distinguished by the strength of the interactions between atoms and molecules depending on the separation among them. Classical electromagnetism plays a big role in atomic and molecular physics and explains their bonding and structure in detail.

This interaction is basically the interaction between charged particles and is based on the basic principles of classical electromagnetism that can later be summarized as Maxwell's equations and are discussed in detail later in this chapter. Electromagnetism deals with the moving charges and the study of accelerated charges indicates that the nonlinear motion of charges can produce a magnetic field that contributes to the electric force perpendicular to the direction of motion of the charge and magnetic field. It means that the electric field (force), magnetic field and direction of motion of the charges are three mutually perpendicular vectors and provide a three-dimensional space for the propagation of electromagnetic waves.

Electrodynamics is the study of time-varying electric and magnetic fields. It involves a detailed study of electromagnetic interaction for different kinds of charge distributions and study of the interaction of current with electric and magnetic fields. The basic laws of electromagnetism known as Gauss's law, Faraday's law and Ampère's law are written together as Maxwell's equations using the electromagnetic wave theory. Maxwell could write the wave equation for electromagnetic fields using these laws and the study of electromagnetic fields and waves is discussed in classical electrodynamics. Propagation of waves through electromagnetic-interaction media is studied in detail accordingly.

4.3.1 Coulomb's law

Electromagnetic force between charges (Coulomb force) is a conservative force and the net work done by this force always vanishes. Charge is an intrinsic property of matter, which is responsible for electromagnetic interaction. Charge has two degrees of freedom; positive and negative charges. One Coulomb charge is defined as a charge which exerts one Newton force on another unit charge when they are at a distance of one meter from each other. This force between two charges is called **Coulomb force**. Independent charge (of an electron) exists in units of electronic charge which is 1.6×10^{-19} coulombs. Direction of the force is parallel or anti-parallel to the line joining the two charges and depends on the polarity of both charges. Magnitude of the Coulomb force is calculated in terms of the charges and the separation between them, however, the work done by this force or the motion resulting due to the application of the force depends on the mass of the affected particles. In other words, the Coulomb force has an impact on the movement of a charged body, which is controlled by Newton's laws of motion. Coulomb force is defined as:

$$\begin{aligned} \vec{F} &\propto \frac{q_1 q_2}{r^2}\hat{r} \\ \vec{F} &= k\frac{q_1 q_2}{r^2}\hat{r} \end{aligned} \tag{4.6}$$

$\hat{r}$ is a unit vector and can be written as $\hat{r} = \frac{\vec{r}}{|\vec{r}|}$. The Coulomb constant $k = \frac{1}{4\pi\varepsilon_0}$ in SI units, where ε_0 is the electric permittivity in free space and is given as $\varepsilon_0 = 8.854\ 287\ 82\ \mathrm{C\ (Nm)^{-2}}$. The speed of light in free space is related to electric permittivity as $c = \frac{1}{\sqrt{(\varepsilon_0 \mu_0)}}$ such that $\mu_0 = 4\pi \times 10^{(-7)}$ newton per square ampere or $\mu_0 = 1.26 \times 10^{(-6)}$ weber per ampere-meter (= newton/square ampere). Weber is the unit of electric flux in SI units defined as tesla per meter square. Magnetic flux is the number of lines of magnetic field per unit area. **Electric field** is a property of charge and is always associated with charge. It can be interpreted as the ability of a charge to exert a force on a unit charge situated at a distance r. An electric field associated with a charge is expressed as a function of distance r and is given in units of newton per coulomb ($\mathrm{N\ C^{-1}}$) as:

$$\vec{E} = k\frac{q}{r^2}\vec{\hat{r}} \tag{4.7}$$

There is another vector field associated with a moving charge called the magnetic field **B** (usually) which actually corresponds to the magnetic flux density. This flux is associated with a moving charge q, exhibiting an accelerated motion which produces a corresponding magnetic force, the **Lorentz Force**. This force is generated by a moving charge perpendicular to the direction of motion of the charge and is given as:

$$\vec{F_B} = \frac{q}{c}(\vec{v} \times \vec{B})$$

The Coulomb force, associated with the electric field, is exerted by the charge, intrinsically, whereas the Lorentz force is created by the moving charges and vanishes as soon as it is put to rest. Therefore, the total electromagnetic force generated by a moving charge is a combination of electric force due to the static charges and the parallel component of the Lorentz force generated by the motion of the charge. This is given as:

$$\vec{F} = q\vec{E} + \frac{q}{c}(\vec{v} \times \vec{B}) \tag{4.8}$$

The velocity $\vec{v}$ in the Lorentz force is a group velocity and corresponds to the velocity of the wave packet. Electric and magnetic fields are usually referred to as classical fields because they are three-dimensional fields and they can acquire any value as long as they satisfy the condition $\vec{E} \perp \vec{v} \perp \vec{B}$ demanded by a vector product. The unit of magnetic field in SI units is the tesla (T), which is equal to newton/(coulomb × meter/sec).

Electromagnetic wave equations for classical fields is associated with apparent wave velocity which is the observable velocity related to the motion of wave from a single point. The propagation of electromagnetic waves in free space takes place by using the constant speed of light which corresponds to the group velocity and is used in obtaining the Lorentz force. and then the group velocity is replaced by c in free space such that $c^2 = \frac{1}{\mu_0 \varepsilon_0}$. Therefore, at the macroscopic scale, the electromagnetic signal speed is always taken as c and the above equations attain the above form.

The electrostatic energy is usually identified as U and is defined as the energy required to move a charge by the electrostatic force to a unit distance r:

$$U = k\frac{q_1 q_2}{r} \tag{4.9}$$

The SI unit of force is the newton and the unit of energy is the joule which can also be expressed as a newton-meter. There is another important quantity in electromagnetism, commonly known as voltage. Voltage is related to the amount of work done by an applied force towards a unit charge. Generally, $u = \vec{F} \cdot \vec{r}$, whereas $V = \vec{E} \cdot \vec{r}$. Electric field is force per unit charge and the voltage is the amount of work done by the electric field to move a unit charge to a unit distance. Voltage can then be written as:

$$V = k\frac{q}{r} \tag{4.10}$$

Another important parameter is current which can be defined as the rate of flow of charge. Current is another important parameter in electromagnetism. It is more like a bulk property of charge indicating the rate of change of the amount of charge per unit time.

4.3.2 Principles of electromagnetic theory

Electromagnetic fields $\vec{E}$ and $\vec{B}$ are continuous fields and are represented as waves. Both of these fields satisfy the wave equations:

$$\left(v_\phi^2 \boldsymbol{\nabla}^2 - \frac{\partial^2}{\partial \mathbf{t}^2}\right)\mathbf{E} = \mathbf{0} \tag{4.11}$$

$$\left(v_\phi^2 \boldsymbol{\nabla}^2 - \frac{\partial^2}{\partial \mathbf{t}^2}\right)\mathbf{B} = \mathbf{0} \tag{4.12}$$

Hence, the electromagnetic fields are considered as continuous variables because they are associated with the flow of charge and their phase changes as a perpendicular component of each other in a plane perpendicular to the propagation vector $\vec{v}$. The change of phase in the electromagnetic fields varies with the movement of individual charges and changes corresponding to the phase velocity v_ϕ. The phase velocity v_ϕ is the velocity of the phase of the electromagnetic waves corresponding to the velocity of the individual state of a wave (every state of the wave packet) and cannot be measured, whereas the group velocity is the overall velocity of the wave and is measurable. Electromagnetic wave equations for classical fields are three-dimensional equations and watch the motion of the individual particles.

This form of wave equation helps to develop a four-dimensional system for electromagnetic waves. By defining four-vectors in the coordinate system $(x, y, z; ct)$, the four-dimensional differential operator de' Alembertian $\Box \equiv \partial_\mu \partial^\mu$ is defined as:

$$\Box = \overrightarrow{\nabla}^2 - \frac{1}{c^2}\frac{\partial^2}{\partial t^2} \tag{4.13}$$

and the three-dimensional coordinate system (x, y, z) takes the four-dimensional form $(x, y, z; ct)$, whereas ct is the positive imaginary coordinate that cannot be expressed in three-dimensional state and three-dimensional momentum operator of quantum mechanics $\left(\hat{p}/_x, \hat{p}/_y, \hat{p}_z\right)$ attains the four-dimensional form $-\hat{p}_x, -\hat{p}_y, -\hat{p}_z; E/c$ and E/c is the imaginary coordinate in this case.

Charge has two degrees of freedom and exhibits force in two forms as attraction and repulsion. Attractive force is directed from a positive to a negative charge, whereas the repulsive force exerted by a charge is moving it away from the other charge in the neighborhood. There are altogether four combinations by two charges and two types (attractive and repulsive) of force. That makes it much more convenient to use the frame of reference of one particle and for most of the calculations, one of the particles is situated at the origin and the displacement vector is defined as separation of charges along the line joining the two charges instead of the spatial coordinates with a randomly chosen frame of reference. Some of the parameters were associated with geometrical arrangement of charge such as dipole moment, quadruple moment and other arrangements.

4.3.3 Gauss's law

Gauss's law for electricity relates the electric field to the total electric charge enclosed by a closed surface. It states that the electric flux through a closed surface is proportional to the total charge enclosed. Electric charge is associated with the electric field and the amount of charge per unit volume is defined as electric charge density. Since we do not observe the magnetic monopoles, the magnetic charge density is always zero. If ρ is defined as free charge density then the variation in the electric field vector is given in terms of ***Gauss's law*** and is written as:

$$\overrightarrow{\nabla}\cdot\overrightarrow{E} = 4\pi\rho \tag{4.14}$$

the corresponding expression for the magnetic field is expressed as:

$$\overrightarrow{\nabla}\cdot\overrightarrow{B} = 0 \tag{4.15}$$

Due to the vanishing of magnetic free charge density magnets are always found as magnetic dipoles, and magnetic monopoles do not exist. However, Gauss's law is defined to be vanishing as we cannot separate both magnetic charges (or magnetic poles). This indicates the major difference between electricity and magnetism. Charge is the intrinsic property of matter, whereas magnetic poles are associated with the motion of charge and the phenomenon of magnetism is associated with the dynamics of charges.

4.3.4 Faraday's law of magnetic induction

A variation in the magnetic field induces a voltage in the circuit. If this variation takes place in a coil, the voltage induced in the coil of wire is called induced EMF that increases or decreases the voltage in the coil. The amount of induced voltage depends on the frequency of variation of the field. This change in voltage depends on the strength of the magnetic field as well as the change in the field.

$$\vec{\nabla}\times\vec{E} = -\frac{1}{c}\frac{\partial\vec{B}}{\partial t} \tag{4.16}$$

This equation describes how a changing magnetic field induces an electric field. It states that the induced EMF around a closed loop is equal to the rate of change of magnetic flux passing through the loop. If a constant electric current passes through a coil, the strength of the magnetic field will depend on the number of turns in the coil. Any change in the magnetic environment of a coil of wire will produce a voltage (induced EMF) in the coil. This change could be produced by changing the magnetic field such as moving a magnet towards or away from the coil, into or out of the magnetic field, rotating the coil relative to the magnet, etc. Faraday's law serves as a succinct summary of the ways a voltage (or EMF) may be generated by a changing magnetic environment. The induced EMF in a coil is equal to the negative of the rate of change of magnetic flux times the number of turns in the coil. It involves the interaction of charge with magnetic field. When an EMF is generated by a change in magnetic flux according to Faraday's Law, the polarity of the induced EMF is such that it produces a current whose magnetic field opposes the change which produces it. The induced magnetic field inside any loop of wire always acts to keep the magnetic flux in the loop constant. If it is decreasing, the induced field acts in the direction of the applied field to try to keep it constant.

4.3.5 Ampère's circuital law

The integral form of Ampère's circuital law for magnetostatics relates the magnetic field perpendicular in a circular wire with the rate of change of electric field and along a closed path to the total current flowing through any surface bounded by that path. This law can be considered as a complementary law to Faraday's law that relates that the variation in electric field and current both contribute to produce some magnetic fields:

$$\vec{\nabla}\times\vec{B} = \frac{4\pi}{c}\vec{J}+\frac{1}{c}\frac{\partial\vec{E}}{\partial t} \tag{4.17}$$

This equation relates the magnetic field to the electric current and the rate of change of electric field. It states that the circulation of the magnetic field around a closed loop is equal to the sum of the electric current passing through the loop and the displacement current, which is proportional to the rate of change of electric field. $\vec{\nabla}\times\vec{B} = \frac{\mu_0}{c}\vec{J}$ ($\vec{J}$ is the current density).

Lenz's law is a magnetism version of Ampère's law. The current induced in a circuit due to a change in a magnetic field is directed to oppose the change in flux and to exert a mechanical force which opposes the motion of charge. $\varepsilon = -N\frac{\partial\phi_B}{\partial t}$ (ε = induced EMF and ϕ_B= magnetic flux).

4.4 Development of classical electrodynamics

It has been mentioned that among the four fundamental interactions known as gravity, electromagnetism, weak and strong interaction, the most well-understood theory is electrodynamics. It is the only fundamental interaction which works both at the microscopic and macroscopic levels simultaneously. Electromagnetism deals with overall behavior of a composite system of charges macroscopically and the flow of charge through neutral objects as current. A detailed study of classical behavior of individual charges or composite structures combining with the wave nature of electromagnetic waves is identified as classical electrodynamics. Electrodynamics is then considered a combined study of interaction of charged particles with radiation, describing it in terms of electromagnetic radiation. Classical electrodynamics is the study of dynamics of charges and describes the interaction of radiation with matter.

Electrodynamics and electromagnetism are similar fields with relatively different classical approaches. Electromagnetism is usually a study of the overall charge of the body and behavior of charge in its totality, whereas electrodynamics deals with the behavior of individual charges. Both of them obey the same rules of electromagnetism and both of these terms are sometimes used interchangeably. However, electrodynamics applies it to individual charges and then integrates the behavior of individual charges to understand the detailed structure of matter. Electromagnetism is a broader concept and deals with the properties of net charges and their relative location to understand the overall behavior of charged objects. It is used to discover laws of electromagnetism to compute the electromagnetic properties of material. Moreover, electromagnetism encompasses the study of electromagnetic phenomena, including static and dynamic electric and magnetic fields, electromagnetic radiation, transmission lines, waveguides, and antennas.

Electrodynamics uses the physics of motion of charges in combination with the electromagnetic radiation which gave rise to Maxwell's equations. It describes the bulk properties of materials integrating the individual properties of charges. However, electromagnetism has a broader scope because it encompasses the study of static and dynamic electric and magnetic fields. It covers electricity and magnetism separately and studies the magnetic properties of materials which has broader application in technology. Electromagnetic phenomena have applications to more charge and current flow and help in creating simple components such as transmission lines, antennas and other basic electric components.

Electrodynamics is used for the detailed study of interacting fluids including plasma and bulk properties of materials in a more precise way. Electrodynamics includes the dynamics of charged particles in electric and magnetic fields, the electromagnetic induction, and the generation of electromagnetic waves. It is extensively used in the study of plasma which can be considered an electromagnetically-interacting fluid with certain density and is electrically neutral. Electrodynamics provides tools to study the electrodynamics of individual charges with quantum mechanics. Electrodynamics deals with electromagnetic fields and waves and translates fundamental laws of electrodynamics using Maxwell's equations.

4.4.1 Maxwell's equations and principles of electrodynamics

The most important fundamental laws of electrodynamics are combined together as Maxwell's equations. These equations were used to develop the electromagnetic wave theory and showed that the laws of electromagnetism satisfy a wave equation. A detailed study of electromagnetic signals was prompted at this point which led to the four-dimensional representation of Maxwell's equations and finally helped in the development of quantum electrodynamics.

We first describe Maxwell's equations in free space where ε_0 and μ_0 give the electric permittivity and magnetic permeability of free space with its constant values. This also gives the constant value to the speed of light as $c = \frac{1}{\sqrt{\epsilon_0 \mu_0}}$ The differential form of Maxwell's equations in free space (in the standard SI units) is written as:

$$\nabla . \mathbf{B} = 0 \tag{4.18}$$

$$\nabla . \mathbf{E} = \frac{\rho}{\varepsilon_0} \tag{4.19}$$

$$\nabla \times \mathbf{E} = -\frac{\partial \mathbf{B}}{\partial t} \tag{4.20}$$

$$\nabla \times \mathbf{B} = \mu_0 \left(\mathbf{J} + \varepsilon_0 \frac{\partial \mathbf{E}}{\partial \mathbf{t}} \right) \tag{4.21}$$

The corresponding integral form of Maxwell's equations in free space (SI units) is written as:

$$\oint \mathbf{E} \cdot \mathbf{dS} = \frac{1}{\varepsilon_0} \iiint_{\Omega} \rho dV \tag{4.22}$$

$$\oint \mathbf{B} \cdot \mathbf{dS} = 0 \tag{4.23}$$

$$\oint \mathbf{E} \cdot \mathbf{dl} = -\frac{d}{dt} \iint \mathbf{B} \cdot \mathbf{dS} \tag{4.24}$$

$$\oint \mathbf{B} \cdot \mathbf{dl} = \mu_0 \iint \mathbf{J} \cdot \mathbf{dS} + \varepsilon_0 \frac{\mathbf{d}}{\mathbf{dt}} \iint_{\Sigma} \mathbf{E} \cdot \mathbf{dS} \tag{4.25}$$

The two systems of units in metric systems are, however, very similar and easily translated in metric notation, but the values of constants are very convenient in a

Gaussian system or CGS (centimeter gram second) units because we can set the Coulomb constant k $= 1/4\pi\varepsilon_0 = 1$ and then the integral form of Maxwell's equations in Gaussian units attains the form:

$$\oint \mathbf{E} \cdot \mathbf{dS} = 4\pi \iiint_{\Omega} \rho dV \tag{4.26}$$

$$\oint \mathbf{B} \cdot \mathbf{dS} = 0 \tag{4.27}$$

$$\oint \mathbf{E} \cdot \mathbf{dl} = -\frac{1}{c}\frac{d}{dt} \iint \mathbf{B} \cdot \mathbf{dS} \tag{4.28}$$

$$\oint \mathbf{B} \cdot \mathbf{dl} = \frac{4\pi}{c} \iint \mathbf{J} \cdot \mathbf{dS} + \frac{1}{c}\frac{d}{dt} \iint \mathbf{E} \cdot \mathbf{dS} \tag{4.29}$$

and the corresponding differential form can be written as:

$$\nabla \cdot \mathbf{B} = 0 \tag{4.30}$$

$$\nabla \cdot \mathbf{E} = 4\pi\rho \tag{4.31}$$

$$\nabla \times \mathbf{E} = -\frac{1}{c}\frac{\partial \mathbf{B}}{\partial t} \tag{4.32}$$

$$\nabla \times \mathbf{B} = \frac{4\pi}{c}\mathbf{J} + \frac{1}{c}\frac{\partial \mathbf{E}}{\partial \mathbf{t}} \tag{4.33}$$

These units become very convenient when we translate Maxwell's equations in a medium and define the displacement vector D corresponding to $D = \varepsilon_0 \varepsilon_r E$ in terms of relative permittivity ε_r. Similarly, the magnetic field in the medium is defined as H and $H = B/\mu_0$. This helps to conveniently define magnetization, polarization and other vectors in a more convenient way and SI units can be retrieved easily, whenever needed. Some of the important units are tabulated, at the end, for comparison and make the transformation of units easy for calculation.

Electric and magnetic fields are usually referred to as classical fields because they are three-dimensional fields and can attain any value. So these fields are considered as continuous variables because they are associated with charges which are composite charges and can be considered continuous variables. This property of electromagnetic fields is a specialty of classical electrodynamics. Maxwell's equations have significant implications that extend beyond their mathematical form. Understanding these equations enables us to comprehend and manipulate electromagnetic phenomena more effectively. Here are some important features of Maxwell's equation which gave birth to electrodynamics which could later be extended to quantum electrodynamics.

- Unifying electricity and magnetism: Maxwell's equations unified the previously separate phenomena of electricity and magnetism, revealing that they are two aspects of the same underlying electromagnetic force. This unification led to important technological advancements, such as the development of electric generators and motors.

- Predicting the existence of electromagnetic waves: From his equations, Maxwell deduced the existence of electromagnetic waves, which travel through space at the speed of light. These waves include radio waves, microwaves, infrared radiation, visible light, ultraviolet radiation, x-rays, and gamma rays. The discovery of electromagnetic waves laid the foundation for wireless communication and the field of optics.
- Quantifying the speed of light: Using his equations, Maxwell was able to calculate the speed of light, which he found to be approximately 299 792 458 meters per second. This precise calculation established the speed of light as a fundamental constant and provided strong evidence for the wave nature of light.

Individual particles have distinct behavior from large masses, especially light ones like electrons. These particles move very fast and acquire relativistic energies that may not be ignorable. Discussion of highly-energetic particles suggests incorporating relativistic contributions at the quantum mechanical scale. For this purpose, the Schrödinger equation of motion has to be updated with relativistic corrections. Maxwell's equations in electrodynamics allow us to describe electromagnetic fields in the form of wave equations. Using the conservation of charge and current, the continuity equation can be written to relate the rate of flow of charge with the divergence of electric current for flowing charges:

$$\nabla \cdot \mathbf{J} + \frac{\partial \mathbf{q}}{\partial \mathbf{t}} = \mathbf{0} \tag{4.34}$$

which is similar to the equation of continuity for fluids that states that the inflow of a liquid in a given volume is equal to the outflow of fluid from the same volume. Another way to say this is then given as:

$$\nabla \cdot (\boldsymbol{\rho}\mathbf{u}) + \frac{\partial \rho}{\partial \mathbf{t}} = \mathbf{0} \tag{4.35}$$

where ρ represents the mass density of fluid and the fluid current is ρu just like the electric current density $\mathbf{J} = \rho v$. This equation shows the conservation of amount of liquid in a unit volume. Similarly, the continuity equation in electrodynamics ensures the conservation of net charge in a unit volume.

The equation of motion for fluids represents the flow of fluid mass and mass density works to determine the number of molecules to a good approximation, whereas for charges the number of particles cannot be determined from the net charge. Polarity of individual particles changes the net charge. On the other hand, the direction and speed of the motion of charge depends on the strength of charges, their polarity and mass. For relativistic systems, spin statistics have to be incorporated that evolve two different equations of motion from Schrödinger's equation for half-integral spin (of fermions) and integral spin (of bosons) as a relativistic theory of quantized fields. Bosons and fermions follow different spin statistics.

In many-particle systems, the equation of motion of charged particles includes spin as well because spin statistics allow bosons to stay in one state with all the same quantum numbers but only one fermion can acquire a particular state at a time. Therefore, the equation of motion of individual charges incorporates spin to determine the right statistics. This difference of spin statistics of fermions and bosons leads to two different equations of motion for bosons and fermions in four-dimensional formalism. This relativistic generalization is not straightforward due to the complexity of mathematics.

4.4.2 Maxwell's equations in four dimensions

Maxwell's equations are the fundamental equations of electromagnetism and they are used to describe the dynamics of charge at non-relativistic energy. Equations (4.18) and (4.19) are three-dimensional forms of Maxwell's equations which are applicable to daily life. These equations work at non-relativistic energies. However, as we know that change of position potentially incorporates the use of time and we add time as an imaginary component, similarly, every charge has the ability to create a magnetic field so we can define a four-dimensional field associated with charge. This field has three real components $\vec{E}$ and one imaginary component $\vec{B} \perp \vec{E}$ as a vector field. Its vector property is associated with the vector nature of the electric field.

The relation between electric and magnetic field is expressed in terms of Maxwell's equations. Equations (4.18)–(4.21) are forms of Gauss's law together give a set of differential forms of Maxwell's equations, and uses all the important laws of electromagnetism that gave us the birth of modern technology. Equation (4.18) leads to

$$\vec{B} = \vec{\nabla} \times \vec{A} \tag{4.36}$$

where A is called a vector potential.

Differential form of Maxwell's equations (4.18)–(4.21) tell us how the variations in the electric and magnetic field depend on each other. The time dependence of magnetic and electric field, in reference to Faraday's law and Ampère's circuital law, relate the time-varying electric and magnetic field to generating electric and magnetic flux. Without getting into detailed discussion of Maxwell's equations, we can look into the corresponding integral form of these equations given in equations (4.27)–(4.30).

Classical electrodynamics works perfectly fine at the macroscopic level just as classical mechanics works perfectly fine at the same scale where we are constrained on the overall behavior of physical systems. Classical mechanics uses Newton's laws of motion for linear systems, the dynamics of orbital motion of rotating systems incorporates Kepler's laws for rotation of bound systems by gravity and fundamental laws of electrodynamics for electromagnetically interacting systems. Gravity and electromagnetic force are both central forces and create gravitationally-interacting orbits in space and electromagnetically interacting orbits in atoms. We just consider dominant force and the angular momenta for the formation of orbits.

The equation of motion of classical mechanics for linear motion is written in terms of the rate of change of linear momentum, whereas the rate of change of angular momentum in circular motion describes the equation of motion for rotation. Net linear force is related to the linear acceleration and net torque on a body is related to the angular momentum. Classical physics deals with continuous variables of a system, in principle, and can attain any value and net force has no restriction on how much acceleration can be produced.

The major difference between gravity and electromagnetism is associated with the basic properties of matter. Mass has only one degree of freedom and gravity is always an attractive force that depends on the quantity of mass and the shape of the objects. Conversely, electric charges have polarity and could be either positive or negative, which gives two types of behavior to the force, either attractive or repulsive. Charged particles still have mass and the mechanical forces are still there along with the electromagnetic force.

Gravity generates orbital motion by competing gravitational attraction with the centrifugal force generated by the angular momentum. Gravitational orbits have no limit on the amount of mass, they just need a large enough force between two masses to ignore any other external force, even gravitation pull of the other objects. Electrodynamics does not produce orbital motion at the macroscopic scale as matter does not carry enough net charge to move large masses measurable distances. In neutral matter, gravity dominates over electromagnetic force at large distances, due to the electrical neutrality of independently existing matter. Neutral matter can only carry partial induced charge, which is not only weak but also temporary. Independently existing neutral matter can be generated by collecting all the positive charges in the center and letting an equal number of negatively charged light particles revolve around it inside the atoms. Therefore, the electromagnetic forces can only generate orbitals of electrons in atoms.

Atoms are electrically neutral and are composed of an equal number of electrons and protons. Protons reside with neutrons in the central part of atom in the nucleus. Neutrality of atoms is indebted to the polarity of charges as an equal amount of charges can balance out to give overall neutral matter. However, the polarity of charge makes it more complicated because opposite charges cancel each other and net charge vanishes. Repulsion between similar charges has to be managed in orbital motion of electrons and the repulsion between protons in the nucleus is controlled by strong interaction between protons and neutrons. Without getting into the details of atomic structure, the nucleus can be considered a single positively charged center and electrons revolve around that center. Almost all the mass of an atom resides in the nucleus while extremely light electrons revolve around it and gravitational interaction is easily ignorable in atoms.

The formation of orbitals of charged particles can occur in the special configuration of atoms only where electrons can revolve around the nucleus but repel one another as well. Moreover, revolving charges exhibit a small magnetic field due to their rotational motion. So electronic orbits not only maintain balance by matching the electromagnetic force and the angular moment, but they also balance the magnetic moment of electrons due to the orbital motion of electrons. This situation is managed by the quantization of angular momentum in atoms, discrete energy levels defined although all of the electrons are identical particles, and considering the size of the nucleus to be small enough to be treated as just a central point.

Electrons are extremely light particles and revolve around the nucleus as well as spin around their own axis of rotation in clockwise or counterclockwise manner. The spin degrees of freedom add correction terms to the orbital angular momentum; these are called spin angular momentum such that the total angular momentum $\vec{J}$ is a vector sum of orbital angular momentum $\vec{L}$ and spin angular momentum $\vec{S}$ which becomes more complicated with the increase in number of particles. Electronic orbitals deal with individual particles so have to incorporate individual particle properties like spin as well. That is the reason that the electron orbits are much more complicated and cannot be described by some laws similar to Kepler's laws of orbital motions (based on central force and angular momentum).

On the other hand, due to the light mass of electrons and tiny size of atoms, classical physics is not enough to describe the electronic motion. Quantum mechanics is needed to understand the motion of electrons inside the atoms. Additionally, electrons move too fast (comparable to the speed of light) to ignore relativity. A detailed study of atomic structure is made possible using quantum mechanics. The key concepts of quantum mechanics are developed using uncertainty principle in the light of wave–particle duality and the operator formalism is instrumented for this purpose. The probabilistic nature of quantum theory, due to the limitations in precise measurements, does not allow us to easily discriminate among different possible states of electrons. The equation of motion of electrons in quantum mechanics is the Schrödinger equation and its solution in spherical polar coordinates leads to the discrete value of angular momenta. This quantization of angular momentum and energy is called the first quantization associated with the quantization of state variables. These quantized variables are identified as quantum numbers and the particles are represented as state functions that are described by the particles as well as wave properties. An overview of quantum mechanics is presented in chapter 1.

Light particles such as electrons, due to small masses, easily acquire relativistic velocities and non-relativistic study may not give the correct information. Even the speed of electrons inside atoms is large but still non-relativistic quantum mechanics works fine. However, electrons can acquire very high velocities at high energies and even other particles may need relativistic treatment. Therefore, special relativity is incorporated into quantum theory, which leads to the second quantization or quantization of fields instead of quantization of variables. Relativistic quantum mechanics evolves into quantum field theory that replaces the first quantization by the second quantization and shows the quantization of fields instead of variables. Therefore, at relativistic energies, particle states are attributed by fields instead of variables. Quantum field theory needs to incorporate spin of the particles in its analysis. However, quantum field theory opens new venues in physics, and acquires a more effective approach to study interaction as a local gauge theory using Lagrangian formalism.

We therefore give an overview of all the relevant theories to identify their scope and then link together all the apparently different approaches in the form of quantum electrodynamics that becomes a standard gauge theory and provides a framework to study all of the fundamental interactions as gauge theories with their inherent properties at the individual particle level. In quantum field theory, the particle spin has to be incorporated and the Schrödinger equation, the equation of motion of quantum mechanics, gives rise to two different equations of motion—namely, the Klein–Gordon equation for particles with integral spin or spin zero and the Dirac equation for half integral spins. The gauge invariance requirement of the Lagrangian led to the discovery of several more quantum numbers.

Chapter 5

Thermodynamics and statistical mechanics

5.1 Introduction

The universe is defined as the sum total of all the matter and energy that we can think about. Matter is identified by its shape, size, color or appearance. It is defined as anything which acquires mass and occupies space, whereas energy is an imaginary quantity and we can only realize through its impact on matter and requirement to do work. Energy is required to apply force to move matter. In this case, mass can be defined as the ability of material objects to resist motion against an applied force, whereas energy is used to change the position of an object in space. We can observe the behavior of matter or the effect of force through experiments. Physics as a subject uses mathematical and computational tools to understand the behavior of matter against the applied force. Mechanical forces are produced in between objects and controlled by operators of forces. These forces can be controlled by various mechanical methods and are used to develop technology.

Modern technologies have emerged from human control on the hidden energy of matter and have the ability to transform one form of energy into another form to use for human benefits. This energy, after conversion into the required form, can produce the required force to perform the required work. Therefore, the study of interaction is relevant for all aspects of scientific and technological development such that the required energy can be created to produce enough force to access the individual particles to study their properties. Force is used to transfer energy in between material objects as well.

A force between material objects can be indirectly realized by the change in its properties and location. Collective properties of matter and its phase may change when it interacts with other matter or goes through the transfer of energy. Scattering allows the change of nature of particles by transferring or sharing energies and momenta and follows certain rules. Collective properties of matter depend on the strength and nature of these interactions and follow certain rules determined by interaction theories and checked by experimental results. The overall energy of

doi:10.1088/978-0-7503-6054-8ch5

macroscopic objects may be an integrated form of all the energies, and external force may be added or subtracted out of it. This macroscopic form of energy of macroscopic objects can change from one form into another form and does not depend on the intrinsic properties of matter only. However, for the integrated form of the constituent particles, energy contributes to different forms of macroscopic energy including kinetic and potential energy. Mass is basically affected by kinetic energy and other properties contribute potential energy.

5.1.1 Bulk properties of materials

Matter is not visible in the form of individual particles and we only see a many-particle system. Individual atoms and molecules exhibit kinetic and potential energy. Average kinetic energy of molecules is defined as temperature T of a system of particles, averaged macroscopic properties or bulk properties of materials include temperature, volume, internal energy and number of particles, etc, and are governed by thermodynamics and their detailed study is managed by statistical mechanics. We briefly discuss both of these theories as they are one of the fundamental approaches to studying physical objects. Thermodynamics overlaps with the field of physical chemistry as these bulk properties handle all the chemical properties to begin with.

A many-particle system exhibits some macroscopic properties such as temperature, pressure, volume, internal energy and some of their special characteristics are defined in terms of the individual particle dynamics of the system. In other words, bulk properties of materials and their overall behavior are expressed as the integral form of individual behavior. Material properties are categorized as intensive and extensive parameters. Extensive properties change with the quantity of matter and depend on mass and volume of the material. The intensive parameters are averaged properties and do not depend on mass and volume or quantity of the material. Temperature and pressure are good examples of the intensive properties, whereas number of particles is an extensive parameter. However, extensive and intensive parameters are related to each other. Therefore, most of the extensive parameters can be redefined as intensive parameters, such as specific heat, specific entropy and molar entropy.

5.1.2 Thermodynamic variables

Three well-known states of matter are called solids, liquids and gases. They are identified by temperature, pressure and specific entropy as intensive parameters. Temperature is defined as the average kinetic energy and pressure is the force exerted per unit area at the surface. Energy, number of particles and volume are examples of extensive properties. In addition to the common parameters such as temperature, pressure, volume, mass and energy, there are certain properties that are specifically defined for the thermodynamic behavior of many-particle systems. Thermodynamic parameters can be defined in terms of the existing parameters. Internal energy of a system, enthalpy and entropy are a few important thermodynamic parameters that are needed to describe statistical properties of a thermodynamic system. Statistical properties are related to the average of individual particle behavior of a system to move in multiple directions and lead to randomness. However, before the discussion of statistical mechanics, we summarize a few relevant concepts and parameters of thermodynamics.

Thermodynamics deals with the dynamics of a bulk material using its bulk properties involving thermal energy. Thermodynamics works with the study of kinetic behavior of a many-particle system and overall properties of bulk material including temperature and other important parameters of the theory. In the case of solids, the thermodynamic quantities of interest may be different than the interesting parameters of fluids. This difference is based on the kinetic behavior of atoms or molecules in a material based on difference of their concentration. Thermal properties of solids are studied as a part of material properties called thermal properties. Thermodynamics provides a better approach to understanding the dynamical properties of fluids due to an increase in degrees of freedom of individual particles. The energy change in fluids is easily measured from their temperature that may cause a change in phase of the material as well.

Temperature is the most conveniently measurable property of a material and is used to define and determine various phases of matter like ice being cold, water having intermediate temperature and vapor temperature being extremely high. Temperature relates to the heat energy of a system and describes the dynamics of the system showing the change in internal energy in terms of the change in temperature. The chemical processes are discussed in terms of the change in temperature or heat energy of a system. The kinetic theory of gases is based on the thermodynamic behavior of gases. However, the study of thermodynamics of systems of particles with some freedom in movement introduces new parameters to describe a many-particle system. We will define a few important parameters here to start with the thermodynamic approach. Bulk properties of matter are used to distinguish between various phases of matter, and density and pressure are the most important parameters. Thermodynamics investigates the concepts of heat and other bulk parameters at the macroscopic level.

5.1.3 Energies in thermodynamics

In addition to the bulk properties such as temperature T, pressure P and volume V, there are some other parameters which are specifically defined for the study of a many-particle system. Temperature is defined as the average kinetic energy and is easily measured and used to differentiate among various thermal systems, whereas pressure and volume are easily measurable bulk parameters as well. Other parameters can be measured in terms of temperature and pressure as well. Some of the other important quantities in thermodynamics are **internal energy** U, **Gibbs free energy** G, **Helmholtz free energy** F and **enthalpy** H of a system. For this purpose, we need to first understand what is meant by **heat** Q. Heat can be understood as the amount of energy needed to move a system of particles due to the individual movement (kinetic energy). Heat is the net energy, whereas temperature is the average energy of the system. **Total energy** of a thermodynamic system is a sum of kinetic energy, potential energy and includes the internal energy such that

$$E = \text{K.E} + \text{P.E} + U$$

The internal energy U is related to other quantities such as enthalpy as well. **Enthalpy** H is a thermodynamic quantity that is used to calculate the heat content Q of a chemical reaction such that it measures the heat flow. The units are usually kJ mol^{-1}. The enthalpy of elemental compounds is zero. It can be considered (another) form of energy and is described as a sum of internal energy and work, whereas **internal energy** is a measure of kinetic and potential energy only. The difference in all various forms of energies corresponds to the nature of a system and the type of work it performs, which leads to a particular change in thermodynamic quantity.

Free energy is needed to perform work by a system. We, therefore, need to define a couple of other commonly used forms of energies for a thermodynamic system such as Helmholtz free energy F and Gibbs free energy G to understand the important forms of usable energies in terms of internal energy and heat. Helmholtz free energy (F) can be defined as: $F = U - TS$. This relation is defined as $\delta Q = \delta W - \delta U$. It corresponds to the working ability of a closed system without changing temperature at all. On the other hand, Gibbs free energy (G) is defined as $G = H - TS$. This way, Gibbs free energy is related to the enthalpy of the system. These parameters can be defined as:

$$F = U - TS$$

$$G = H - TS$$

$$H = U + PV.$$

5.2 Laws of thermodynamics

Thermodynamics discusses the comparison of temperature and an average kinetic energy of a many-body system. Thermodynamics deals with the transfer of heat or management of the temperature between two systems. Matter can allow the transfer of thermal energy (heat) via conduction, convection or radiation. Movement of particles from one place to another is not possible in solids but particle transfer is possible in fluids (liquids or gases). Therefore, we do not need to worry about the change in composition of connected solids. Fluids on the other hand allow the transformation of molecules among themselves when they are connected to each other through an opening which may allow the transition of particles among themselves. At this point we need to define thermodynamic variables such as temperature, pressure, density, chemical potential, internal energy, entropy and so on. The flow of thermal energy between two systems or the transfer of temperature is understood by thermodynamics and is based on a few basic principles.

When two systems are in thermal contact with each other, they are said to acquire thermal equilibrium when they reach the same temperature and further flow of heat stops. Flow of heat and movement of particles from one system to another system takes place through the flow of heat in the form of kinetic energy of the particles in fluids. Two systems do not allow the flow of heat between them if they are in thermal equilibrium. Two systems at thermal equilibrium have the same temperature and do not allow the transfer of heat. Thermodynamic equilibrium means that two thermodynamic systems have all the matching parameters where further transfer of energy between two systems is not possible.

There are four fundamental laws of thermodynamics. These laws describe the relationships among different thermodynamic parameters such as temperature, internal energy, and entropy of the system. These laws are used to study the total work done by a thermodynamic system including the work done by the heat and other thermodynamic parameters. Two systems are defined to be in thermal equilibrium if there is no net flow of heat between them.

Every physical many-body system is called a thermodynamic system when it can be defined in terms of its bulk properties including temperature, density, volume, mass and chemical composition, etc. All these properties contribute to the thermodynamic potential. The flow of material between two thermodynamic systems takes place from high potential to low potential along with thermodynamic parameters. They are all related to the fact that the heat flow into or out of a thermodynamic system affects the internal energy of the system according to the law of conservation of energy.

5.2.1 Zeroth law of thermodynamics

If two mutually independent systems A and B are in thermal equilibrium with each other, and system A is in thermal equilibrium with a third system C, then C will also be in thermal equilibrium with B. This is called the zeroth law of thermodynamics. It means that if two systems are in thermal equilibrium with each other and one of the systems is in equilibrium with a third system, then all three systems are in thermal equilibrium. It is equivalent to say if two bodies are in thermal equilibrium with a third body, then they are said to be in equilibrium with each other.

5.2.2 First law of thermodynamics

If two systems are in thermal contact with each other, the flow of energy can take place but the total amount of energy between two systems remain unchanged. The second law of thermodynamics can be considered as another form of law of conservation of energy which is relevant in thermodynamics. In other words, the law of conservation of energy between thermally connected systems, are put in isolation.

5.2.3 Second law of thermodynamics

This law describes the mechanism of flow of heat without breaking any other law of physics. The second law simply states that the flow of heat is always directed from a hotter to a colder region unless this flow is interrupted by any other stronger force. Spontaneous flow of heat from a hotter to a colder region can more easily be related to the entropy of the system. However, all of the heat cannot be transformed into energy to perform work.

It can also be described in terms of heat energy of a system. If an amount of heat is added to a system, that heat is either used by the system or it is released out of the system in one manner or another. It means that the change in thermal energy of a system can be accounted for in terms of the internal energy δU and the work done by a system called δW. We can write:

$$\delta E = \delta Q - \delta W$$

where Q represents heat of the system and W represents the work done by the system or work on the system. δx_i gives the measurable change in thermodynamic variables during a thermodynamic process.

The second law of thermodynamics ensures the **entropies** of two interacting systems manage the flow of heat until a thermodynamic equilibrium is established and the entropy of the two systems reaches the optimum value and stops the flow of heat. It just provides the limitation of a thermodynamic behavior that it cannot be fully reversible. Or in other words, when an equilibrium is established between two systems, a quantity called entropy is maximized. Entropy S is defined as a measure of disorder and the flow of heat Q can be related to change in entropy as:

$$\delta S = \frac{\delta Q}{\delta T}.$$

The rate of change of entropy is directly related to the change of temperature. As soon as the entropy reaches the maximum disorder (equilibrium), the flow of heat stops which prevents the irreversible processes after an isolated system of two small systems acquire an equilibrium. So, the flow of heat is not possible as soon as a system reaches the condition:

$$\delta S \geq \frac{\delta Q}{\delta T}.$$

This is the condition which never lets the heat flow in a backward direction and prevents reversibility. Entropy helps to define another law of thermodynamics. However, we need to develop a better understanding of entropy before getting into the third law of thermodynamics.

5.2.4 Entropy

Entropy is another quantity that is introduced in thermodynamics as well. It is a physical parameter that can be considered as a measure of intrinsic disorder in a material. It gives the amount of energy that is used to produce randomness and is not available for the mechanical work on individual particles. Entropy can be defined in reference to certain measurements as Shannon entropy, Neumann entropy and so on. Entropy is constant at absolute zero only, which is an ideal situation that is physically impossible. For example, the entropy of a perfect crystal at zero temperature (zero Kelvin) is zero.

Entropy is represented by S and keeps on changing in active materials and can be considered as a measure of the dispersal of energy. It can be considered as a thermodynamics quantity that measures the disorder in a closed system, which is related to the randomness and uncertainty of a thermodynamic system. Entropy of an isolated system remains constant for reversible processes. However, the combined entropy of systems and their environment must increase for irreversible processes. It is directly related to the microstates. Increase in a microstate is related to the increase in entropy and increased entropy cannot decrease automatically.

As soon as we understand the concept of entropy, we can redefine second law of thermodynamics in terms of entropy as the entropy is always increasing and cannot be decreased spontaneously. This means that heat cannot flow from a colder to a hotter system spontaneously. However, a closed system can stay at its maximum value of entropy or at the same temperature until it is in contact with a colder (hotter) system.

5.2.5 Third law of thermodynamics

The third law of thermodynamics is based on the fact that every chemical or physical process is associated with a change in entropy. It can sometimes be related to the Nernst theorem as the entropy of a system approaches a constant value at absolute zero temperature and is related to the nature of a system as:

$$\lim_{T \to 0} \Delta S = 0$$

It shows that the entropy at equilibrium reaches a constant value.

$$\lim_{T \to 0} (\text{System in Equilibrium}) = 0$$

The third law sometimes can simply be stated as the entropy of a closed system always tends to approach its maximum value. This maximum value is obtained at equilibrium. Whereas its entropy approaches a minimum value as soon as the temperature of a system approaches zero.

5.2.6 *TdS* equations

A set of important equations of thermodynamics using various forms of energy are called TdS equations relating temperature T and the related change in entropy dS keeping various other thermodynamic quantities constant (one at a time). These TdS equations are derived from the following initial relations:

$$\begin{aligned} TdS &= dH - VdP \\ TdS &= dE = dQ - PdV \end{aligned} \tag{5.1}$$

$$TdS = dU + PdV \tag{5.2}$$

A set of more useful relations using thermodynamic parameters also called TdS equations can be derived from equation (5.1). For this purpose, we define $\beta = \frac{1}{T}$ and $\kappa = \frac{1}{T}(\frac{\partial T}{\partial P})_v$

$$TdS = C_v dT + T\left(\frac{\partial P}{\partial T}\right)_v dv = C_v dT + \frac{\beta T}{\kappa} dv$$
$$TdS = C_p dT - T\left(\frac{\partial v}{\partial T}\right)_p dP = C_p dT - \beta v T dP \tag{5.3}$$
$$TdS = C_p\left(\frac{\partial T}{\partial v}\right)_p dv + C_v\left(\frac{\partial T}{\partial P}\right)_v dP = \frac{C_v \kappa}{\beta} dP + \frac{C_p}{\beta v} dV$$

C_v and C_p are specific heats at constant volume and specific heat at constant pressure, respectively.

$$C_p = \frac{(TdS + PdV)}{dT} = T\left(\frac{\partial S}{\partial T}\right)_p + P\left(\frac{\partial V}{\partial T}\right)_p$$

Specific heat can be defined as the amount of heat required to increase the temperature of a unit mass of material measured in grams by 1 Kelvin such that $C_p > C_v$ always.

5.2.7 Kinetic theory of gases

The kinetic theory of gases is a simple classical model for the study of the thermodynamic behavior of gases. Gases are considered as non-interacting multi-particle systems. This basic version is associated with non-interacting gases and gives a relationship between temperature T, pressure P, and volume V. The proportionality relation among T, P and V is given as the ideal gas law such that:

$$\frac{PV}{T} = \text{constant.}$$

It also relates other macroscopic parameters or bulk properties of gases. Transport properties of gases in terms of the transport parameters such as viscosity, thermal conductivity and diffusion are studied.

1 gram-mole of an element or a compound is expressed in terms of its atomic weight or molecular weight expressed in grams, respectively. 1 mole has $6.022\ 1415 \times 10^{23}$ particles and this number is called Avogadro's number. Even a small amount of material is a collection of a very large number of particles and statistical treatment is the only way to study the internal behavior of the system. Thermodynamics gives an overall behavior of a multiparticle system derived from a statistical approach. Most of the thermodynamic quantities are averaged quantities. A few important thermodynamic quantities are introduced above and will be understood in terms of statistical variables in detail later. For a better understanding of statistical mechanics, we need to define and understand commonly used parameters of treating many-body systems using the probability theory of statistics.

5.3 Introduction to statistical mechanics

Dynamics of single-body or two-body problems can be understood by simple mathematical techniques. As soon as we come across a many-particle system, we have to treat it using the concept of probability and statistical mechanics. The ideal gas treatment of a non-interacting system is pretty simple and can be applied to a large system with a countless number of non-interacting particles. The kinetic theory of gases is used for an ideal gas where atoms and molecules exhibit random motion and have sufficiently low density. It is a system of a very large number of particles moving freely in a large space of a gaseous system. Phase is determined by standard definition of various phases and properties of material.

In statistical mechanics we deal with particles that are chemically identical and where all of them can stay in the same state together which can create a bulk of material in solid, liquid or gaseous states. These particles may be distinguishable due to some other properties such as momenta and energies. However, they can randomly move in three-dimensional space. Due to this random motion we can treat a non-interacting gas as a random-walk problem analogous to random motion in a system of non-interacting identical particles. It is known in chemistry that a physical system composed of many identical particles seems to exhibit random motion described as **Brownian motion**, which is random motion in a fluid.

Precision and predictability are basic needs of science and single-particle behavior at microscopic level is predictable and precise measurements are possible to an acceptable level if the initial conditions are known. However, the predictability is replaced by probability when more than one outcome is possible and a system is not physically accessible. Many-particle systems have several possibilities in their mutual behavior and the impact of the applied force may not be fully predictable due to some missing information or due to too many possible outcomes. Probability theory works when there is some information that cannot be accessed due to extremely small objects or a very large number of samples and the missing information is incorporated by averaging over all the possible theoretical or experimental outcomes based on the known and tested laws of the corresponding theory.

Before getting involved in the study of statistical behavior of matter we need to define a few important statistical terms such as **statistical ensemble**, canonical ensemble, grand canonical ensemble and other statistical terms. **Ensemble** is a collection (group) of a large number of particles that share their individual properties in a system. A thermodynamic system is defined as a statistical ensemble of a very large number of particles in a closed system. It is defined for a collection of material with uniform composition where all the particles are identical and behave in the same way. In statistical mechanics a statistical ensemble is a complete set of states that may be occupied by identical particles. Sometimes, it may be described as a set of possible states for the same particle. Certain conditions can identify this set of states differently. A statistical ensemble is an infinite number of identical particles which are indistinguishable and exhibit random motion for a non-interacting system which can be treated as virtual copies of the same system (or multiple copies of the same particle with all the same parameters). **Thermodynamic ensembles** are a special type of systems that are in statistical equilibrium among themselves. Therefore, the statistical properties of a system can be derived using classical or quantum mechanics. Out of the general definition of ensemble, specific cases can be derived.

A **canonical ensemble** is an ensemble which is recognized by a large number of similar particles, and this system may weakly be connected to a heat bath.

The energy of such a system may not be fully known. However, the temperature is specified for such a system and the system is a fully closed system in a way that particles cannot move in or out of the system. However, in a **grand canonical ensemble** even the number of particles is not constant. It can be described as an open system which is in thermal contact with a reservoir which allows the flow of energy, number of particles, and the associated change due to electrical contact, radiative contact, chemical or thermal contact and so on. An **ensemble average** is an averaged quantity which is an average of probability of existence of microstates because a statistical ensemble could be defined locally.

5.3.1 Entropy

Entropy is another important quantity which is introduced in thermodynamics but it acquires various forms in statistical mechanics. It is a measure of intrinsic disorder within a closed system. It gives the amount of energy that is used to produce randomness and that thus is not available for mechanical work on intrinsic particles. Entropy can be defined in reference to certain measurements as Shannon entropy, Neumann entropy and so on. Entropy is statistically constant at absolute zero and it keeps on changing in active materials. Various states in a physical system can be described in terms of n number of discrete states which can be converted into a continuous set of states. However, reversibility and conservation of information can be required. Statistical mechanics deals with probabilities and is used differently at the quantum scale and the states of the particles are expressed in terms of probabilities. Therefore, entropy becomes a parameter related to all possible numbers of states. In an identical particle system, the entropy is used in the form of probability of various quantum states and the entropy is described in terms of quantum numbers.

5.3.2 Probability

Probability theory in statistical mechanics considers all states with the possibility of finding a particle in a certain state. All of these states can be equally probable or not. Probability theory is one of the key concepts of statistics, and it is used exclusively in statistical physics to be able to deal with a large number of states with a sufficiently large number of particles. The probability of finding an object is related to the availability of a state, which depends on the distribution of various states. The concept of probability is very simple if we have enough information about the available state and all states are equally probable. The example of rolling a six-sided cubic dice is an appropriate example of an unbiased probability. If a dice has six equally probable sides, then the probability of getting a specific side (identified by a particular number) is equal to the number of observed sides (which is always 1 at a time) is divided by the total number of states, giving $P = \frac{1}{6}$. We can translate this concept of probability into statistical mechanics. We can have biased or unbiased distributions of states. The most general definition of unbiased probability for n particles can then be found by defining a probability of existence of a state with n particles defined $\mathcal{N}$ such that:

$$\mathcal{N} \equiv \frac{\text{number of particles } n}{\text{total number of states}} = \frac{n}{N}$$

for randomly distributed N states where every state is equally available to each and every particle and only one particle can be found in one state at a time. However, this concept of simple discrete probability becomes much more complicated when the states become biased or incoming particles have their own priorities.

We can easily distinguish between the discrete and continuous distribution of states. The above definition corresponds to discrete states and can be easily evaluated. Total number of states can then be expressed as ($N = \sum_{i=1}^{N} i$) for discrete states and for continuous distribution of states ($\rho = \int_1^{\rho} dx$). Then the probability of finding a continuous set of states $f(x)$ in this region ρ is written as:

$$p(\rho) = \int_1^{\rho} f(x)dx.$$

If the probability distribution of a state i, ρ_i, can contribute to the total distribution of probability it is normalized as ($\sum_i \rho_i = 1$). If we look for a state with maximum entropy at thermodynamic equilibrium:

$$S = -k\sum_i \rho_i ln\rho_i.$$

Total number of available microstates can be defined in terms of energies of the individual states E_i with $i = 1, 2, 3, \ldots$. These states make a complete set of states $\Omega(E)$ to apply statistical mechanics properly. Most of the important parameters of a thermodynamic system such as the density of states, average energy of the system, partition function and entropy are related to the number of states and partition function.

5.3.3 Maxwell's relations

Maxwell's relations are derived from the conservation of the second derivatives of thermodynamic potential with respect to two natural variables of thermodynamics like temperature T, Pressure P, volume V and entropy S. All other parameters such as the coefficient of thermal expansion α, compressiblity κ, heat capacity at constant volume C_V and heat capacity at constant pressure C_P are imported from thermodynamics as it is. It uses the general form of Schwarz's theorem for any two variables i and j which are not necessarily thermodynamic variables such that:

$$\frac{\partial}{\partial x_j}\left(\frac{\partial \Phi}{\partial x_i}\right) = \frac{\partial}{\partial x_i}\left(\frac{\partial \Phi}{\partial x_j}\right)$$

where Φ is a function including a statistical function and x_i and x_j are statistical variables. Maxwell's relations in thermodynamics are derived using functional forms of internal energy $U = U(S, V)$, enthalpy $H = H(S, P)$, Helmholtz free energy $F = F(T, V)$ and Gibbs free energy $G = G(T, P)$. These relations are derivable from the symmetry of second derivatives using the definition of thermodynamics potential. It is also noticeable that all thermal variables are defined in terms of simple relations among various thermal variables, whereas the statistical parameters are a simultaneous function of various statistical variables. It is therefore important to discuss the rate of change of one parameter as a partial derivative of other parameters keeping everything else constant.

$$\left(\frac{\partial T}{\partial V}\right)_S = \left(-\frac{\partial P}{\partial S}\right)_V = \left(\frac{\partial^2 U}{\partial V \partial S}\right)$$

$$\left(\frac{\partial T}{\partial P}\right)_S = \left(\frac{\partial V}{\partial S}\right)_P = \left(\frac{\partial^2 H}{\partial S \partial P}\right)$$

$$\left(\frac{\partial S}{\partial V}\right)_T = \left(\frac{\partial P}{\partial T}\right)_V = \left(-\frac{\partial^2 F}{\partial T \partial V}\right)$$

$$\left(-\frac{\partial S}{\partial P}\right)_T = \left(\frac{\partial V}{\partial T}\right)_P = \left(\frac{\partial^2 G}{\partial T \partial P}\right)$$

All of these equations can be rewritten using the properties of partial derivatives:

$$\left(\frac{\partial y}{\partial x}\right)_z = 1\Big/\left(\frac{\partial x}{\partial y}\right)_z$$

where all of the statistical relations are evaluated in terms of partial derivatives instead of a simple change in thermodynamic parameters.

5.3.4 Gibbs free energy

All the concepts of statistical parameters are related to quantum mechanical concepts. We can relate all these parameters in terms of quantum mechanical operators. The Gibbs entropy of a macroscopic classical system is a function of probability distribution over the phase space of an ensemble and can be written in terms of the partition function as:

$$S = -\lambda_2 U + k ln Z,$$

whereas:

$$\frac{dS}{dU} = -\lambda_2 \equiv \frac{1}{T}$$

Calculation of probability in statistical mechanics is not so straightforward because it depends on how the possible states depend on the properties of particles and their behavior as a many-particle system. Therefore, statistical mechanics goes way beyond the simple calculation of probability. We need to define several statistical functions in terms of statistical properties which are defined in terms of the available state of the system.

5.3.5 Relation between statistical mechanics and thermodynamics

Mathematical development of statistical mechanics is based on the generalization of thermodynamic concepts. The system of identical particles is expressed in terms of distribution of velocities or in other words, their kinetic energies. Statistical mechanics generalizes the concept of energies for statistical systems. Energy is used as a parameter to distinguish between various identical particle states and various quantum numbers are used to distinguish between energies. Entropy, density of states and the particle distribution $f(x)$ are all related to the number of available states with all identical parameters.

5.3.6 Partition function

Partition function is used to describe thermodynamic variables as statistical quantities in terms of the set of states. A partition function is constructed incorporating the properties of a statistical ensemble. Therefore, partition functions correspond to the properties of an ensemble, which could be a canonical ensemble, a grand canonical ensemble, or even a micro-canonical ensemble.

$$Z = \sum_i e^{-\beta E_i} = \sum_E \Omega(E) e^{-\beta E}$$

So Z can be related to the number of states $\Omega(E)$ of a micro-canonical ensemble between energies E and $E + \delta E$ and their average over energy is related to energy E, where δE is much smaller than the size of the corresponding canonical ensemble. It is related to the density of states, such that:

$$\Omega(E) = \mathscr{W}(E)\delta E,$$

where $\mathscr{W}(E)$ is called the density of states. The probability distributions ρ_i and entropy S are then expressed as:

$$\rho_i = \frac{1}{Z} e^{-\beta E_i}$$

and finally, the entropy is defined as derivatives of entropy keeping one or the other parameter constant such as:

$$\begin{aligned} dS &= \frac{dQ}{T} = \frac{dE + pdV}{T} \\ \left(\frac{\partial S}{\partial E}\right)_{V,N} &= \frac{1}{T} \\ \left(\frac{\partial S}{\partial V}\right)_{E,N} &= \frac{p}{T} \end{aligned} \tag{5.4}$$

We could relate the entropy with Z such that:

$$S \equiv k(lnZ + \beta\overline{E})$$

or using

$$S \equiv \frac{U}{T} + klnZ$$

we obtain:

$$\begin{aligned} dS &= \frac{1}{T}dE + \frac{p}{T}dV - \sum_i \frac{\mu_i}{T} \\ dE &= TdS - pdV + -\sum_i \mu_i dN_i \end{aligned} \tag{5.5}$$

Other parameters such as entropy S and chemical potential μ of the theory are expressed in terms of the partition function. Partition functions are linked with the number of states of a system and most of the physical quantities can be defined in terms of partition functions:

$$P_i = \frac{e^{-\beta E_i}}{\sum_i e^{-\beta E_i}} \tag{5.6}$$

$$\mu_i = \left(\frac{\partial E}{\partial N_i}\right)_{S,\,V,\,N} = \left(\frac{\partial F}{\partial N_i}\right)_{T,\,V,\,N} = \left(\frac{\partial G}{\partial N_i}\right)_{T,\,P,\,N} \tag{5.7}$$

such that:

$$dG \equiv d(E - TS + pV) = -sdT + Vdp + \sum_i \mu_i dN_i \tag{5.8}$$

Another way to define chemical potential is:

$$\mu = kTln\frac{N_x}{Vqx/\lambda_x^3}$$

where G is the Gibbs free energy and F is the Helmholtz free energy. Using the functional forms of energy is to understand dependence of various energy variables to the parameters of the system (depending on the type of ensemble, under consideration) as:

$$\begin{aligned} S &= S(E,\,V,\,N_1,\,N_2,\,N_3,\,\ldots) \\ G &= G(T,\,P,\,N_1,\,N_2,\,N_3,\,\ldots) \\ E &= E(S,\,V,\,N_1,\,N_2,\,N_3,\,\ldots) \\ F &= F(T,\,V,\,N_1,\,N_2,\,N_3,\,\ldots) \end{aligned} \tag{5.9}$$

Applications of various equilibrium conditions, based on the phase of the system corresponding to the type of ensemble, helps to describe various types of energy with basic statistical variables. They help us evaluate various parameters of every system in equilibrium. This is how equilibrium statistical mechanics works. Application of non-equilibrium statistical mechanics to a physical system is much more challenging technically and numerical methods have to be used usually to solve problems of non-equilibrium physics for some valid approximations, according to the requirement of information.

Various distribution functions are expressed in terms of various properties of states depending on quantum statistics. The **Maxwell–Boltzmann** distribution function f is given as:

$$f(x) \propto e^{\frac{-x^2}{2a^2}}$$

with a positive constant 'a' and x can be any distribution variable. It can be normalized by a factor $2a\sqrt{\frac{2}{\pi}}$, giving:

$$f(x) = 2a\sqrt{\frac{2}{\pi}}e^{\frac{-x^2}{2a^2}}.$$

In a canonical ensemble, the probability distribution in 'i' number of states can be given as $e^{\beta E_i}$, and the average energy of the system is defined as:

$$\overline{E} = -\frac{\sum_i e^{\beta E_i} E_i}{\sum_i e^{-\beta E_i}} = \frac{1}{Z}\frac{\partial lnZ}{\partial \beta} \tag{5.10}$$

where $Z = \sum_i e^{-\beta E_i}$ and $E_i = \sum_i e^{-\frac{\partial}{\partial\beta}}Z$. So the average of a measurable parameter is calculated in terms of all the states of a system. So the canonical distribution is given by:

$$P_i = \frac{e^{-\beta E_i}}{\sum_i e^{-\beta E_i}}$$

whereas the **partition function** is defined as a statistical function that describes the statistical properties of a physical system while it stays in thermodynamic equilibrium. 'i' corresponds to a state in an ensemble and is given by:

$$Z = \sum_i e^{-\beta(n_1\varepsilon_1 + n_2\varepsilon_2 + \cdots)}$$

The upper limit of 'i' will be the maximum number of available orthogonal states in a system. The total number of possible ways the N particles can be distributed in available states is $\frac{N!}{n_1!n_2!n_3!\cdots}$ The variable β is a statistical variable defined as $\beta = \frac{1}{kT}$. $k = 1.380649 \times 10^{-23}$ J K^{-1} is the Boltzmann constant. It is also expressed as $k = 8.617\ 333\ 262 \times 10^{-5}$ eV K^{-1}. $e^{-\beta E_i}$ is called the Boltzmann factor, otherwise.

5.4 Introduction to quantum statistical mechanics

Quantum mechanics and statistical mechanics are both based on probability theory to understand the dynamical behavior of matter. Single-particle quantum mechanics looks at the probability of the presence of that particle in a set of a certain number of equally probable states which is called an ensemble. However, in statistical mechanics we deal with a many-particle system and the particle states are identified by their energies. Therefore, quantum mechanics can be applied to statistical mechanics to study the dynamics of many-particle systems using quantum mechanics. The effect of spin and other quantum numbers on particle dynamics can be incorporated to distinguish between the particles' behavior.

A straightforward generalization of kinetic theory of gases gives the statistical mechanics of bosons or spinless particles. Non-interacting gas particles can be dealt with as spin zero particles (bosons) in quantum mechanics. Quantum statistical mechanics is a generalization of a statistical approach to quantum mechanics and finds a way to include fermions along with bosons. Quantum mechanics has a distinct feature of quantization of angular momentum and includes spin to identify a group of identical particles which exhibit different dynamical behavior due to different spin.

A complete partition function for bosons can be written as:

$$Z = \sum_{n_1, n_2, \cdots} \frac{N!}{n_1! n_2! n_3! \cdots} e^{-\beta(n_1\varepsilon_1 + n_2\varepsilon_2 + \cdots)}$$

Quantum mechanics then has another class of particles called fermions with an antisymmetric wavefunction. Fermions have to obey the Pauli exclusion principle, which states that none of the quantum mechanical states can let two particles reside simultaneously unless they have opposite spins (spin up versus spin down). Two fermions with opposite spins can share the same orbital or energy level. Therefore, quantum statistical mechanics differs from the classical version of classical mechanics in that it can be applied to a system of particles which can be distinguishable due to their spin. So due to the presence of distinguishable particle collections, we can define new parameters to apply **quantum statistical mechanics**.

In quantum statistical mechanics, we define a statistical ensemble by a density operator which is a non-negative self-adjoint operator in Hilbert space *H*, which describes the quantum mechanical system with fermionic or bosonic states. The concept of a quantum mechanical system is discussed in detail in a separate chapter but the difference in the behavior of quantum states due to spin is identified as spin statistics. Bose statistics allows any number of particles to occupy the same state such that all the atoms and molecules can be found with the same energy and its partition function reads as:

$$Z = \left(\sum_i e^{-\beta\varepsilon_i}\right)^N = \frac{1}{(1 - e^{\beta\varepsilon})}$$

In this partition function the individual particle energies ε_i are classically summed and average energy is calculated for the system. Quantum mechanically, all these energies follow the distribution of energies using quantum statistics and it leads to the generalization of velocities for the Maxwell distribution to incorporate Fermi statistics and Boson statistics and ends up getting two different particle distribution functions for fermions and bosons. At this point we have to mathematically develop quantum statistical physics for physical applications.

5.4.1 Formulation of quantum statistical physics

Particle states are defined by $\psi_i(x_a)$, which represents an ith particle with x_a coordinates. Fermi statistics indicates that two particles can be represented by an antisymmetric combination of two particles and gives a negative sign if both particles are interchanged such that the two particle states can get the following combination of two states as:

$$\psi_i(x_1)\psi_i(x_2) - \psi_i(x_2)\psi_i(x_1).$$

On the other hand, boson statistics can lead to two particle states where interchange of particles has no effect on the combined state and is mathematically represented as:

$$\psi_i(x_1)\psi_i(x_2) + \psi_i(x_2)\psi_i(x_1).$$

Obviously, the difference between fermions and bosons is distinguishable when we have more than one particle. It becomes more relevant for multiparticle systems and the density of states will be determined by the spin statistics.

For a many-particle system, if we have n_i as the number of particles in the ith state with energy ε_i and $\mathcal{N}$ as the total number of particles of a system:

$$\mathcal{N} = \sum_{i=1}^{k} n_i$$

and the total energy of $\mathcal{N}$ particles system is given as:

$$E = \sum_{i=1}^{k} n_i\varepsilon_i = n_1\varepsilon_1 + n_2\varepsilon_2 + \cdots + n_k\varepsilon_k.$$

In order to calculate statistical parameters such as the average number of particles in the ith state with energy ε_i, we use:

$$\bar{n}_i = -\frac{1}{\beta}\frac{\partial lnZ}{\partial \varepsilon_i} = \frac{\sum n_i e^{-\beta n_i \varepsilon_i}}{\sum e^{-\beta n_i \varepsilon_i}}$$

This leads to the total number of particles as:

$$\mathcal{N} = \sum_{0}^{\infty} e^{-\beta n_i \varepsilon_i} \tag{5.11}$$

and the particle distribution for the bosons is the Bose–Einstein distribution of bosons, given as:

$$\bar{n} = = \frac{e^{-\beta\varepsilon}}{(1 - e^{-\beta\varepsilon})} = \frac{1}{(e^{\beta\varepsilon} - 1)} \tag{5.12}$$

This is called the **Planck distribution**. The Bose–Einstein distribution corresponds to the partition function:

$$Z = e^{-\beta(n_1\varepsilon_1 + n_1\varepsilon_2 + n_2\varepsilon_1 + \cdots + n_i\varepsilon_i)}$$

for the same state i number of particles allowed in quantum states due to quantum statistics. We discuss the statistical formulation of various statistical ensembles and their mathematical formulation. Various statistical ensembles are defined by another hypothetical source of heat (thermal reservoir) called a heat bath.

5.4.1.1 The canonical ensemble

A canonical ensemble is composed of two thermal systems (in mutual equilibrium) which are in contact with a heat bath as well. In this case, the two systems are able to exchange energy while maintaining a constant temperature. Meanwhile the exchange of particles or the change in volume is not allowed. P_i is the probability at a given time t, that we will find the system to be in a state characterized by energy value E_i. Such a system is a member of a canonical ensemble which is defined by (N, V, T) where $\mathcal{N}$ identical systems $\{n_i\}$, constitute the ensemble that shares an energy $\mathcal{E}$.

$$\sum_i n_i = \mathcal{N}$$
$$\sum_i n_i E_i = \mathcal{E} = \mathcal{N} U$$

where U is the average energy per system in the ensemble. A set of possible states $\{n_i\}$ satisfies the above conditions in a possible mode of distribution of the total energy $\mathcal{E}$ among the $\mathcal{N}$ members of the ensemble. Each mode can be realized in various ways with distinct permutations, denoting it with the symbol $W\{n_i\}$:

$$W\{n_i\} = \frac{\mathcal{N}!}{n_0!n_1!n_2!\cdots} = \frac{\mathcal{N}!}{\prod_r (n_i)!}$$

for an ensemble with equally probable conditions. In this case, the frequency at which the distribution set $\{n_i\}$ appears is directly proportional to $W\{n_i\}$. Therefore, W is maximized for the most probable mode of distribution. Probability of a canonical distribution can be found as:

$$P_i \equiv \frac{\langle n_i \rangle}{\mathcal{N}} = \frac{e^{-\beta E_i}}{\sum\limits_r e^{-\beta E_i}}$$

The average energy is then given by:

$$U = \frac{\sum_r E_i e^{-\beta E_i}}{\sum r e^{-\beta E_i}} = -\frac{\partial}{\partial \beta} ln\left\{\sum_i e^{-\beta E_i}\right\}$$

Now, we look to extract the information of the macroscopic properties of a given system to relate to our statistical formulation. Using Helmholtz free energy, we find the following relation:

$$U = A + TS = A - T\left(\frac{\partial A}{\partial T}\right)_{N,\,V} = -T^2\left[\frac{\partial}{\partial T}\left(\frac{A}{T}\right)\right]_{N,\,V}$$
$$= \left[\frac{\partial (A\,/\,T)}{\partial (1\,/\,T)}\right]_{N,\,V} = \left[\frac{\partial}{\partial \beta}\left(\frac{A}{kT}\right)\right]_{N,\,V}$$

Here we find a close relation between the quantities in the statistical formulation and the thermodynamic relations $\beta = \frac{1}{kT}$:

$$ln\left\{\sum_i e^{-\beta E_i}\right\} = -\frac{A}{kT}.$$

This gives us the most fundamental result of the canonical ensemble theory, written in the form:

$$A(N, V, T) = -kTln\ Z_N(V, T)$$

where:

$$\mathcal{Q}_N(V, T) = \sum_i e^{-\beta E_i}$$

The quantity $\mathcal{Q}_N(V, T)$ is known as the partition function. The partition function is also known as the sum-over-states. While the dependence on T is explicitly seen, the dependence on N and V comes from the energy eigenvalues E_i. The partition function is a particularly important result, as it relates the macroscopic thermodynamic quantities of a system to its microscopic details.

5.4.1.2 The grand canonical ensemble

A grand canonical ensemble is a further generalization of the canonical ensemble. In this ensemble, particles are allowed to move in between two systems. However, we use the same approach as we used in the mathematical description of a canonical ensemble and try to develop a partition function for a grand canonical ensemble. If $n_{i,j}$ at a time t corresponds to a system with N_i particles and E_j energy, then the set of numbers $\{n_{i,j}\}$ represents a possible mode of distribution of energy E_j and particles N_i among the $\mathcal{N}$ members of the ensemble. This set must follow the following conditions:

$$\sum_{i,j} n_{i,j} = \mathcal{N}$$
$$\sum_{i,j} n_{i,j} N_i = \mathcal{N}\langle N \rangle$$
$$\sum_{i,j} n_{i,j} E_j = \mathcal{N}\langle E \rangle$$

At this point, we use the most probable mode of distribution in the canonical ensemble, which correspond to distinct permutations, such that:

$$W\{n_{i,j}\} = \frac{\mathcal{N}!}{\prod_i (n_{i,j})}$$

However, the parameters α and β will be determined using the equations which calculate average quantities in a given ensemble and its partition function:

$$\frac{\langle n_{i,j} \rangle}{\mathcal{N}} = \frac{e^{-\alpha N_i - \beta E_j}}{\sum_i e^{-\alpha N_i - \beta E_j}}$$

$$\langle N \rangle = \frac{\sum_{i,j} N_i e^{-\alpha N_j - \beta E_j}}{\sum_{i,j} e^{-\alpha N_i - \beta E_j}} \equiv -\frac{\partial}{\partial \alpha}\left\{\ln \sum_{i,j} e^{-\alpha N_i - \beta E_j}\right\}$$

$$\langle E \rangle = \frac{\sum_{i,j} E_s e^{-\alpha N_i - \beta E_j}}{\sum_{i,j} e^{-\alpha N_i - \beta E_j}} \equiv -\frac{\partial}{\partial \beta}\left\{\ln \sum_{i,j} e^{-\alpha N_i - \beta E_j}\right\}$$

We can now establish a connection between the grand canonical ensemble and thermodynamics by defining a parameter q. By taking a derivative of q, and using the previously derived equations for averages of $\langle N \rangle$ and $\langle E \rangle$, as well as the most probable mode of distribution, we can take the differential of q:

$$dq = -\langle N \rangle d\alpha - \langle E \rangle d\beta - \frac{\beta}{\mathcal{N}} \sum_{i,j} \langle n_{i,j} \rangle dE_i$$

$$d(q + \alpha\langle N \rangle + \beta\langle E \rangle) = \beta\left(\frac{\alpha}{\beta} d\langle N \rangle + d\langle E \rangle - \frac{1}{\mathcal{N}} \sum_{i,j} \langle n_{i,j} dE_j \rangle \right)$$

We can interpret this by comparing it to the statement of the first law of thermodynamics:

$$\delta Q = d\langle E \rangle + \delta W - \mu d\langle N \rangle$$

This gives us the following result:

$$\delta W = -\frac{1}{\mathcal{N}} \Sigma_{r,s} \langle n_{r,s} \rangle dE_s, \quad \mu = -\alpha/\beta$$

$$d(q + \alpha\langle N \rangle + \beta\langle E \rangle) = \beta\delta Q$$

Since β is obtained by integrating over the change in heat δQ, then β must be inversely proportional to absolute temperature T and $\beta = 1/kT$ with k, the Boltzmann constant and $\alpha = -\mu/kT$ where μ is the chemical potential. Given that $\delta Q = TdS$, then we can solve for q such that:

$$q = \beta TS - \alpha\langle N \rangle - \beta\langle E \rangle$$

$$q = \frac{TS + \mu\langle N \rangle - \langle E \rangle}{kT}$$

The term $\mu\langle N \rangle$ is equal to the Gibbs free energy of the system, which is given by $G = \langle E \rangle - TS + PV$:

$$q \equiv \left\{ \ln \sum_{i,j} e^{-\alpha N_i - \beta E_j} \right\} = \frac{PV}{kT}$$

This is the essential link between the statistics of the grand canonical ensemble and the thermodynamics of the system, a relationship with central importance to the formalism. Now, we introduce a parameter z, defined as the fugacity of the system and given by:

$$z \equiv e^{-\alpha} = e^{\mu/kT}$$

The q-potential now takes the form:

$$q \equiv \left\{ \ln \sum_{r,s} z^{N_r} e^{-\beta E_s} \right\} = \left\{ \ln \sum_{i,j} z^{N_i} Q_{N_i}(V, T) \right\} \quad (\text{with } Q_0 \equiv 1)$$

Now, defining a new parameter $\mathcal{Q}$ as the natural logarithm of the q-potential, we now have:

$$q(z, V, T) \equiv \ln \mathcal{Q}(z, V, T)$$

thus:

$$\mathcal{Q}(z, V, T) \equiv \ln\mathcal{Q}(z, V, T) = \sum_{r, s} z^{N_i} Z_{N_i}(V, T) \quad (\text{with } Q_0 \equiv 1)$$

This parameter $\mathcal{Q}$ is the grand partition function.

5.4.2 Application of quantum statistical physics

The partition function is the most important quantity to study the statistical properties of a material. All the statistical parameters can be evaluated from the partition function of a system which carries all the information about the statistical distribution of energy among the particles in various states. The simplest definition of classical probability is a ratio of favorable conditions divided by the total number of possibilities as:

$$\frac{\text{favorable states}}{\text{total states}} \tag{5.13}$$

These states could indicate anything including an event, a process, or an arrangement or any occurrence starting from a rolling dice to a single particle representing as a quantum state.

The average number of particles in a state can be calculated from the dispersion relation obtained by the partition function Z. This dispersion relation is given as:

$$\overline{n}_i = \frac{\text{Number of states with particular energy}}{\text{Total number of states in an ensemble}} = \frac{e^{\beta\varepsilon_i}}{Z}$$

$$\overline{n}_i = -\frac{1}{\beta}\frac{dlnZ}{ln\varepsilon_i} \tag{5.14}$$

Using this relation (in equation (5.14)) for the dispersion relation, we can find the distribution functions for various kinds of particles. Later on, various thermodynamic parameters which contribute to particle energies can be calculated using the appropriate partition function. We can use the partition function to evaluate the averages energy of the system for a given ensemble by averaging it over the partition function. Starting with the calculation of average energy of the system as:

$$\overline{E} = -\frac{\partial lnZ}{\partial\beta} \tag{5.15}$$

and

$$\overline{E}^2 = -kT\left(\frac{1}{Z}\frac{\partial^2 Z}{\partial\beta^2}\right) \tag{5.16}$$

We can calculate the average parameters in a statistical ensemble. In general:

$$\overline{X} = -kT\frac{\partial \ln Z}{\partial x} \tag{5.17}$$

such that the average pressure can be calculated as:

$$\bar{p} = -kT\frac{\partial \ln Z}{\partial V} \tag{5.18}$$

and the chemical potential is written as:

$$\mu_i = \frac{\partial F}{\partial N_i} = -kT\left(\frac{\partial \ln Z}{\partial N_i}\right)_{T,\ V_i} \tag{5.19}$$

5.4.3 Quantum statistics of ideal gases

Kinetic theory of ideal gas gives equal distribution of velocities of particles in each direction, which is related to the temperature, pressure and volume of the gas, depending on the number of particles.

5.4.3.1 Maxwell–Boltzmann statistics

Classical distribution of an ideal gas of N identical particles is represented by the partition function involving a sum of n_r for all R states:

$$Z = \sum_r e^{-\beta(n_1\varepsilon_1 + n_2\varepsilon_2 + n_3\varepsilon_3 + \cdots)}$$

where for a total of N molecules, the number of permutations is:

$$^{r}C_n = \frac{N!}{n_1!n_2!n_3!\cdots}.$$

The average number of particles in energy state ε_i can be calculated as:

$$\bar{n}_i = -\frac{1}{\beta}\frac{\partial lnZ}{\partial \varepsilon_i} = -\frac{1}{\beta}N\frac{-\beta e^{-\beta\varepsilon_i}}{\sum_i e^{-\beta\varepsilon_i}}$$

which gives the Maxwell–Bolzmann distribution. It is worth noticing that the limit $\beta \to \infty$ leads to $\bar{n}_e \to N$. If temperature tends to zero kelvin, an infinite number of particles can stay in the same energy state. This indicates a multiparticle classical system which does not put any restriction on the number of particles in a state. On the other hand, it can be easily seen that for $\beta \to 0$ or for large values of T the number of particles is proportional to N.

The second moment of n or dispersion of number of particles is given by:

$$(\Delta\bar{n}_i)^2 = \sum_i^N P(n_i)(n_i - \bar{n}_i)^2 = -\frac{1}{\beta}\frac{\partial \bar{n}_i}{\partial \varepsilon_i}$$

Quantum statistics requires one to include the intrinsic spin of particles. Fermions with a half-integral spin obey Fermi statistics and spinless scalars and bosons with integral spins obey boson statistics which does not put any restriction on the number of particles in any state. The photon, as a vector boson with spin one, is a massless particle and is quantum mechanically treated as a wave as well as a particle. Its statistics can be discussed individually.

5.4.3.2 Photon statistics

For photons, the following functions are defined $Z = (\sum_1^\infty e^{-\beta\varepsilon_i})^N$ which leads to:

$$\overline{n}_i^{\gamma} = \frac{1}{e^{\beta n_i} - 1}$$

which goes to infinity because in the limit of large T where $\beta \to 0$, an infinite number of particles can be found at any temperature. The only difference between photons and other bosons is that they always have zero chemical potential because photons are massless.

5.4.3.3 Boson statistics

For a general boson with nonzero mass, the above relations can be generalized by including another parameter α such that:

$$Z = \left(\sum_0^\infty e^{-(\alpha+\beta\varepsilon_i)} \right)^N$$

which leads to:

$$\overline{n}_i^{B} = \frac{1}{e^{(\alpha+\beta\varepsilon_i)} - 1}$$

which goes to infinity because in the limit of large T where $\beta \to 0$, an infinite number of particles can be found at any temperature. The parameter α can be evaluated using the relation:

$$\alpha = -\beta \frac{\overline{\partial F}}{\partial N} = -\beta \mu_i$$

where μ_i is the chemical potential, which depends on the number of particles and mass. It is defined as the energy needed to remove a particle from a system. Chemical potential is obviously zero for photons. The photon itself is a quanta of energy.

5.4.3.4 Fermion statistics

Fermions obey the Pauli exclusion principle due to their half-integral spin and more than one particle cannot stay in the same quantum state. However, due to the half-integral spin, a spin up particle can stay in the same state as a spin down particle. In this case, the particle distribution depends on whether another particle is already there in a state or not and the partition function is written as:

$$Z = \left(\sum_1^\infty e^{-(\alpha+\beta\varepsilon_i)} \right)^N$$

which leads to the particle distribution function of fermions as:

$$\overline{n}_i^{F} = \frac{1}{e^{(\alpha+\beta\varepsilon_i)} + 1}$$

which goes to infinity because in the limit of large T where $\beta \rightarrow 0$, an infinite number of particles can be found at any temperature. The parameter α can be evaluated using the relation:

$$\alpha = -\beta \frac{\overline{\partial F}}{\partial N} = -\beta \mu_i$$

where μ_i is the chemical potential of fermions and is related to the number of particles through their mass.

IOP Publishing

Samina S Masood

Chapter 6

Quantum mechanics

6.1 Introduction

Quantum mechanics provides the most suitable framework for a detailed investigation of atomic and molecular structures considering all of the available options about the states of individual particles in a system. For this purpose, the concept of probability is employed from statistical mechanics. In other words, quantum mechanics and statistical mechanics commonly use probability to study the dynamics of non-interacting individual particles. However, a statistical mechanics approach does not go beyond the classical scale, whereas quantum mechanics works well on smaller scales. It defines a connection between probabilities and relates it to the compact form of realities. Probabilities are associated with states which may be considered as waves or particles and when probability collapses into reality, it creates observable particles with definite measurable realities.

Quantum mechanics deals with very tiny compact objects at the molecular and atomic scale. It can also be applied at a subatomic scale and size. It provides tools for a detailed study of atomic and molecular structures. Detailed understanding of the atomic structure and molecular formation along with the ability to do the precise calculation of energies of electrons and atoms can explain the probability and preferences among various chemical processes. Atoms, as part of molecules or by themselves, can release or gain electrons and exhibit an overall positive or negative charge and produce positive and negative ions, respectively. This happens during chemical reactions and these charged atomic or molecular components are called ions or radicals and facilitate chemical reactions through ionic bonding.

Quantum mechanics is constructed on the basis of the uncertainty principle, which is introduced to mechanics along with the concept of wave–particle duality to incorporate the probabilistic nature into quantum mechanics. Particles are then identified as states and exhibit particle and wave properties simultaneously. These

doi:10.1088/978-0-7503-6054-8ch6

states are indicated by wavefunctions which carry intrinsic properties such as mass, charge and other physical properties of particles, yet they are still identified by wave properties such as wavelength and frequency as well. The concept of a de Broglie wave is associated with matter waves when particles in motion acquire wave properties due to periodic motion. These waves are expressed in terms of frequency and the wavelength λ such that the magnitude of the three-momentum p is $\mathbf{p} = |\mathbf{p}|$. These matter waves are indicated as state functions $|\psi\rangle$ which can simultaneously behave as a particle and a wave and the state functions are described in terms of particle and wave properties.

The quantum mechanical study of tiny objects at the individual particle level indicates that another parameter is associated with charged particles to explain the stability of atoms and explain why charged electrons in atoms are not accelerated due to circular motion and do not cause atomic disintegration or the emission of radiation. It must explain the unsolved mystery of atomic stability. Particles can be classified as either fermions or spin half (or half integral spin) particles, or they can be classified as bosons that have spin zero (or integral spin).

The identification of fermions is very important as they have antisymmetric wavefunctions in contrast to boson states which are symmetric in nature. The antisymmetric nature of the wavefunction gives an explanation of the stability of atoms in terms of the Pauli exclusion principle. This principle states that two fermions with exactly the same properties (quantum numbers) can reside in the same state, which explains why electrons are distributed as pairs with opposite spins in atoms, justifying the stability of atoms. This also defines the orthogonality of quantum states. Atoms are then composed of mutually orthogonal states of quantum mechanics.

Atoms are bound together through electromagnetic interactions. A detailed study of atomic structure is made possible using quantum mechanics. The key concepts of quantum mechanics are developed using uncertainty principle in the light of wave–particle duality and the operator formalism is instrumented to determine the parameters of quantum states. The probabilistic nature of quantum theory and limitations in precise measurements do not allow discrimination among different possible states of electrons. The Schrödinger equation is used as the equation of motion of electrons in atoms and its solution in spherical polar coordinates for electromagnetic potential leads to the quantization of angular momentum and energy, called the first quantization. These quantized variables are identified as quantum numbers and the particles are represented as state functions that are described by the particles as well as wave properties. The properties of states are determined by the operators and the eigenvalues associated with these operators.

However, orbital formation takes place at the individual particle level in atoms with the help of the Pauli exclusion principle and electrodynamics. Atoms are electrically neutral and are composed of an equal number of electrons and protons. Neutrality of atoms is indebted to the polarity of charges as an equal amount of charges can balance out to give overall neutral matter. However, the polarity of charge makes it more complicated because opposite charges cancel out the net charge but similar charges repel each other as well. The formation of orbitals of charged particles can occur in the special configuration of atoms only where electrons can revolve around the nucleus but maintain certain distance due to Coulomb repulsion. Moreover, revolving charges have an associated magnetic field as well. The electronic orbits not only balance the electromagnetic force and angular moment but also take care of the magnetic field generated by the orbital motion of electrons.

A detailed structure of atoms and the dynamics of particles at subatomic level has to be studied in terms of quantum mechanics, which is primarily based on the uncertainty principle or wave–particle duality. Quantum mechanics works on the basis of probability theory and prefers to incorporate all possible states of a system in this study. Therefore, the particles are represented as state functions that simultaneously use the particle and wave properties. However, the probability may compromise on precision in finding the exact location and describe a system as a linear combination of various states along with the corresponding probabilities.

On the other hand, the microscopic form of electromagnetic interaction has the same behavior for individual charges at small scales. Quantum mechanics deals with dynamics of individual electrons inside atoms and molecules. However, simultaneous applications of relativity, quantum mechanics and electrodynamics, without ignoring the basic principles of classical physics, shape up quantum mechanics with its distinct features. Such a complete theory which fully describes the electromagnetic interaction of relativistic systems at quantum scale is called quantum electrodynamics (QED), which is a gauge theory and can be treated as quantum field theory of an electromagnetically-interacting system.

6.2 A brief overview of quantum mechanics

Quantum mechanics is a theory of microscopic systems, which is based on the uncertainty principle due to the wave–particle duality. All particle states are considered to exhibit wave properties such as mass and charge and the wave properties as frequencies and wavelength simultaneously. The particle momenta and energies are related to wavelengths and frequencies using the concept of de Broglie waves. Every particle in quantum scale is described as a wave, as well.

Wave–particle duality does not allow simultaneous measurement of position and momentum precisely. This unusual behavior of individual particles is called the uncertainty principle, which leads to the quantization of momentum and hence the theory is labeled as quantum mechanics. The uncertainty principle makes it difficult to precisely identify the state of a particle as a wavefunction that carries properties of waves and particles. These states are labeled in terms of state variables like position, momentum and energy. Quantum mechanics cannot distinguish between identical particles and identifies them by their intrinsic properties such as mass, charge and spin. Identical particle states are distinguished by state variables such as momentum, energy and spin, etc, and are distinguished by the set of state variables. This is because more than one state could exist with the same values for a few or more parameters, but they can be distinguished by even one differing parameter. Therefore, the uncertainty principle as a manifestation of probability theory and quantum mechanics keeps all the probable states as orthogonal states so that if a particle acquires any of these states, it implies that it cannot be found in any other state.

Particles are identified by state functions or wavefunctions of incoming state ψ and its conjugate matrix is represented as a row vector or an outgoing state $\psi^{\dagger}$. All of these state functions are labeled by state variables. This wavefunction ψ completely describes a state of the system and bears all the properties of particles and waves. Some of these properties are not even recognized in classical mechanics such as individual particle spin and parity. It is therefore called the state function. It does not even have any physical interpretation unless the incoming and outgoing states are both found to be the complex conjugate of each other, in the effort to track the correct transition. This is mainly because we cannot be certain about the existence of a state at a particular position or its momentum as the position and momentum cannot be measured simultaneously due to the uncertainty principle.

The state functions are defined as vectors in the corresponding vector space. Components of the wavefunction are complex numbers. If the incoming state is a column vector with complex components the outgoing state is represented as a row vector with all the complex conjugates of incoming state components. We define a state ψ as a linear combination of several mutually independent state vectors such that the state of the system can be written as $\psi = a_1\psi_1 + a_2\psi_2 + \cdots$. The probability of finding a particle in a state ψ is defined as:

$$P = |a_1\psi_1|^2 + |a_2\psi_2|^2 + \cdots$$

and all $|a_n|^2$ correspond with the probability of each corresponding state $|\psi_n|^2$ such that $|a_1|^2 + |a_2|^2 + \cdots = 1$. Normalization of probability is a requirement for the existence of a physical system and we can only find the relative probability of different states after normalizing the wavefunction such that total probability is calculated as:

$$P = \langle\psi|\psi\rangle = \int \psi(x)^{\dagger}\psi(x)dx = \sum_n \langle\psi_n|\psi_n\rangle = 1$$

In the matrix notation, the probability can be expressed as a product of a row vector and a column vector as:

$$P = \begin{bmatrix} \psi_1^* & \psi_2^* & \cdot & \cdot & \cdot \end{bmatrix} \begin{bmatrix} \psi_1 \\ \Psi_2 \\ \cdot \\ \cdot \\ \cdot \end{bmatrix}$$

The uncertainty principle and the probabilistic behavior due to limitations in measurement led to establishing an operator formalism for quantum mechanics. The properties of state functions are checked individually by operating an operator on the state functions. These operators are matrix operators and give the information about the state. When an operator operates on a state and gives a precise value without changing the state itself, the operator is called an eigenvalue operator, which can identify the properties of a state without modifying the state itself. Any equation represented as a general operator $\hat{O}$, gives an eigenvalue λ. $\hat{O}$ is called an eigenvalue operator and ψ is called an eigenfunction. Quantum mechanically, the measurable eigenvalues are real. They could be the only measurable value of the corresponding property of a state. The general form of the eigenvalue equation is written as:

$$\hat{O}|\psi\rangle = \lambda|\psi\rangle \tag{6.1}$$

where $|\psi\rangle$ is the incoming state in the above equation. The corresponding eigenvalue equation for the outgoing state $\psi^\dagger$ can be written as:

$$\langle\psi|\hat{O}^\dagger = \lambda^*\langle\psi| \tag{6.2}$$

and the eigenvalue λ can only be measured for a Hermitian operator $\hat{O}^\dagger = \hat{O}$ that has a real eigenvalue $\lambda = \lambda^*$, and the symbol $\dagger$ corresponds to the complex conjugate and the transpose of the operator $\hat{O}$.

The normalization of a state is required for the correct information of the physical state. For a wavefunction corresponding to an incoming state $|\psi(x)\rangle$ as a function of position x, the physical interpretation of the wavefunction requires information about the outgoing state at the same point $\langle\psi(x)|$ as well. The total probability P of the existence of a state for any value anywhere in space is given as:

$$P = \langle\psi(x)|\psi(x)\rangle = \int \psi(x)^\dagger\psi(x)d^3x \tag{6.3}$$

and it is normalized to unity to find the distribution of probability among different states. The average or mean value of an operator is calculated as

$$\langle\hat{O}\rangle = \frac{\int \psi^\dagger\hat{O}\psi d^3x}{\int \psi^\dagger\psi d^3x} \tag{6.4}$$

The eigenvalue equation of the Hamiltonian operator is called the Schrödinger equation. The most general form of the Schrödinger equation is given in terms of the Hamiltonian H. The equation of motion in quantum mechanics is written in terms of energy. Total energy of a system is calculated to figure out the state of a system and its behavior. Classically, total energy of a system is expressed in terms of the Hamiltonian $H = T + V$ where T is the kinetic energy and V is the potential energy of the system. Exact measurements of position and momentum makes it possible to separate the kinetic energy and potential energy from each other. Quantum mechanically, this measurement is not possible due to wave–particle duality and the uncertainty principle. Therefore, the operator formalism is used. The Hamiltonian is an operator and is operated on a state giving the total energy of the system in the corresponding state. This is represented by the Schrödinger equation as an eigenvalue equation such that $\hat{H}$ is an eigenvalue operator giving a measurable energy E as the real eigenvalue of the operator $\hat{H}$ in state ψ:

$$\hat{H}|\psi\rangle = E|\psi\rangle \tag{6.5}$$

The Hermitian operator $\hat{H}^{\dagger} = \hat{H}$ corresponds to the energy operator that can only find out the energy of the state ψ with the real eigenvalue of energy E that is physically measured for a normalized state. Quantum mechanics tells us that the eigenvalue of a Hermitian operator is real, meaning that hermiticity is required for the operators that correspond to physically measurable quantities. The general form of the Hamiltonian $\hat{H}$ is written in terms of kinetic and potential energy of a system giving $\hat{H} = \frac{\hat{p}^2}{2m} + V$ and the general form of the Schrödinger equation in non-relativistic quantum mechanics attains the form:

$$\left(\frac{\hat{p}^2}{2m} + \hat{V}\right)\psi(x, t) = E\psi(x, t) \tag{6.6}$$

Quantum mechanics deals with wave–particle duality and can explain the wave behavior of a particle, considering them as de Broglie waves of matter particles such that the particle waves are written as:

$$\lambda = \frac{h}{p}$$

in the usual notation where λ corresponds to the wavelength and p the magnitude of momentum. Planck's constant h is the scale of quantization in quantum mechanics and is used to express the uncertainty in the measurement of conjugate variables. First quantization in quantum mechanics expresses some of the state variables as an integral multiple of h. It is actually a unit in phase space and is associated with the uncertainty between a pair of conjugate variables such as energy and time and the position and momentum. On the other hand, electromagnetic light waves are described in terms of massless quanta of their energy called photons, which could be treated as particles as well.

The initial application of quantum mechanics and its development is associated with atomic and subatomic physics. A detailed study of the hydrogen atom was performed by solving the Schrödinger equation in three-dimensional space. This study was not possible without developing specialized mathematical tools, but it described the quantization of angular momentum in the bound electromagnetic system and the existence of discrete orbitals in atoms and discrete energy states. Energy associated with different orbitals of atoms was calculated in terms of the Bohr radius as:

$$E_n = \frac{n^2h^2}{8\pi^2a^2} \tag{6.7}$$

where a corresponds to the Bohr radius of the hydrogen atom, n is the state index and E_n gives the energy of the nth state.

The probability of transition of quantum states from one state into another state can be represented as

$$\langle\psi'(x)|\psi(x)\rangle,$$

which indicates the transition of state $|\psi(x)\rangle$ to $\langle\psi'(x)|$. Since

$$\sum_n |\phi_n(x)\rangle\langle\phi_n(x)| = 1,$$

a complete set of intermediate states can be introduced as a path of transition from the initial to the final state such that:

$$\sum_n \langle\psi'(x)|\phi_n(x)\rangle\langle\phi_n(x)|\psi(x)\rangle.$$

The basic equation of motion of quantum mechanics is called the Schrödinger equation, given by:

$$\hat{H}|\psi\rangle = E|\psi\rangle$$

which can be expressed in detail as:

$$\left(\frac{-\hat{p}^2}{2m} + \hat{V}\right)|\psi > = E|\psi\rangle \tag{6.8}$$

The Schrödinger equation of motion is the main equation of motion and is used to study all quantum mechanical systems in different coordinates with the relevant boundary conditions and helps to evaluate the behavior of the state functions in the given potential under the given physical conditions. However, the laws of motion of classical mechanics do not work directly because the particle's position becomes uncertain due to its wave behavior. Also, the exact dynamical behavior is described in terms of one of the possible states. Therefore, the state function in quantum mechanics represents the probability of finding a particle in that state and not the exact values of all the relevant parameters.

The probabilistic nature of quantum mechanics leads to the development of an operator formalism in quantum mechanics to find the most probable values of those parameters. In the operator formalism, the component of x corresponds to $\hat{p}_x = -\iota\hbar\frac{\partial}{\partial x}$, which can be written in the form of a vector operator as $\vec{p} = -\iota\hbar\hat{\nabla}$. The the total energy operator, the Hamiltonian, can be written as:

$$\hat{H} = \frac{-\hat{p}^2}{2m} + \hat{V}.$$

In quantum mechanics, all of the state variables are defined as operators. The real measurable eigenvalues are obtained by the hermitian operator only. Therefore, the Hamiltonian needs to be a hermitian operator to give a physically measurable value of energy as the eigenvalue.

The Lagrangian of classical mechanics is translated into a quantum mechanical operator to evaluate the action of force and is expressed in terms of quantum mechanical operators. The total energy of the system is also expressed as the

Hamiltonian operator such that the Lagrangian and the Hamiltonian of the system are written as:

$$\begin{aligned} H &= T + V = \frac{-p^2}{2m} + V \\ L &= T - V = \frac{-p^2}{2m} - V \end{aligned} \tag{6.9}$$

Almost all the physical states in quantum mechanics can be discussed by solving the Schrödinger equation using the correct potential and defining the relevant boundaries.

6.3 Examples of quantum mechanical systems

A few examples of quantum mechanical calculations are discussed here just to give an idea of how quantum mechanics could be applied to physical systems for very simple cases of constant potential. The standard problems of quantum mechanics treat bound states as potential wells and forbidden states as potential barriers. A few well-known cases are:

- Infinite square well potential: This potential may provide a permanent trap and a particle can never move out of it.
- Finite square well potential: This potential shows a bound particle state inside a potential box. A particle cannot come out of this box as long as the particle's energy cannot overcome the box potential.
- Step potential: This infinite potential barrier corresponds to a totally forbidden state.
- Finite potential barrier and tunneling. that is used to calculate reflection, refraction and transmission probabilities of electromagnetic signals. Leakage is also understood using this type of potential.
- Scattering is not described by a single potential. However, scattering can produce a pair of particle and antiparticle which can be represented as the creation and annihilation of the particle, respectively. Deep inelastic scatterings processes may identify the structure of the target particles as well. It will be discussed in the next chapters.

We discuss the above cases with constant potential V_0 to begin with. However, these potentials are not constant for actual physical systems. So the solution of Schrödinger equations for all the above cases depend on the nature of potential. We discuss the simple one-dimensional Schrödinger equation in Cartesian coordinates only. In actual physical situations, the three-dimensional Schrödinger equations should be solved using the proper form of the potential with the spatial variation and time. In many cases, we may need spherical potentials and the three-dimensional differential equations can be solved using the techniques described in the second chapter. Here we just demonstrate the application of boundary conditions to solve the Schrödinger equation for simple one-dimensional forms of each of the above potentials. These techniques lead to a much more involved mathematical approach for actual physical systems.

6.3.1 Potential well

The potential well represents a potential that is lower inside the boundaries as compared to the outside potential. So it could either be constructed with positive or negative potentials. We are considering square-shaped wells in one-dimensional space in Cartesian coordinates. A positive potential can be a potential well between for the higher outside potentials, usually taken as infinite walls. The potential well with positive outside potentials is similar to a classical box that has a particle trapped inside and the trapped particle cannot come out. If this box is made up of infinite boundaries, it is called an infinite potential box and a particle inside the box can never come out and anything from outside the box cannot enter due to the infinite boundaries. In this case, the infinite potential walls of the well are similar to classical rigid walls. A particle in such a box under a constant potential exhibits sinusoidal waves, creating standing waves in different modes.

The most common potential well represents a bound state with negative potential. Considering the simplest case of constant potential, we can write a square well potential, as shown in figure 6.1:

$$V = -V_0, \quad \text{for} \; -a \leq x \leq a$$

$$V = 0, \quad \text{otherwise}$$

The wavefunction ψ_1, ψ_2, ψ_3 in various regions of the square well are given as:

$$\begin{aligned} \psi_1 &= A_1 e^{k'x} + A_2 e^{-k'x} \\ \psi_2 &= B_1 e^{ikx} + B_2 e^{-ikx} \\ \psi_3 &= C_1 e^{k'x} + C_2 e^{-k'x} \end{aligned} \tag{6.10}$$

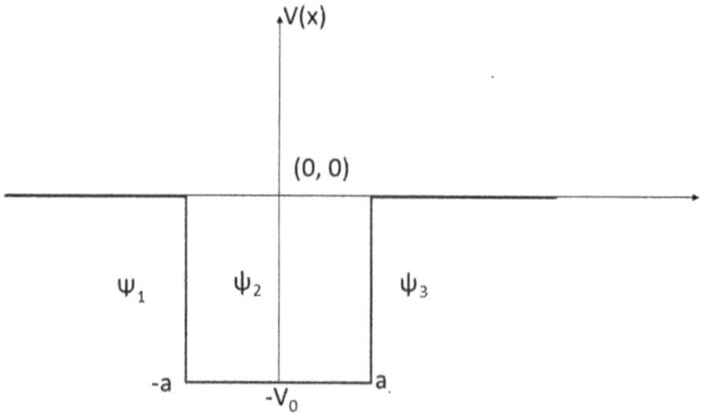

Figure 6.1. Potential well for a constant negative potential.

with:

$$k^2 = \frac{2m(E + V_0)}{\hbar^2}, \text{ and } k'^2 = -\frac{2mE}{\hbar^2} \tag{6.11}$$

for $E < 0$. The corresponding boundary conditions (B.C.) are given as:

$$\begin{aligned} \psi_1(-a) &= \psi_2(-a) \\ \psi_2(a) &= \psi_3(a) \\ \psi'_1(-a) &= \psi'_2(-a) \\ \psi'_2(a) &= \psi'_3(a) \end{aligned} \tag{6.12}$$

Using equation (6.11) in equation (6.12), we can obtain with a little algebra:

$$\begin{aligned} \psi_1 &= A_1 e^{k'x} \\ \psi_2 &= B_1 e^{\iota kx} + B_2 e^{-\iota kx} \\ \psi_3 &= C_2 e^{-k'x} \end{aligned} \tag{6.13}$$

for $\psi \to 0$ when $x \to \pm\infty$. The two possible solutions of the above equations are:

$$\begin{aligned} (ka)\tan(ka) &= (k'a) \\ (ka)^2 + (k'a)^2 &= \left(\frac{2mV_0}{\hbar^2}\right) \end{aligned} \tag{6.14}$$

The required solution may be formed for the relevant potential in a suitable coordinate system.

The potential well with finite potential walls holds the particle inside, and requires higher energy than the barrier potential to move the particle out of the well. So the walls of the potential well act like a barrier for a particle inside the well. The stability of bound states depends on the depth of the barrier. Barrier depth represents the stability of the bound states, whereas the height of the potential barrier indicates the energy required to cross the barrier. Atoms are just like a potential well for electrons and ionization potential is similar to barrier potential.

6.3.2 Potential barrier

Bound states of particles behave as potential barrier. An energy greater than V_0 is needed for particles to cross the barrier. The infinitely high potential barrier corresponds to classical rigid walls which can never be crossed. For the potential barrier, width of the barrier is also important to study the behavior inside the barrier. A very general example of potential barrier is in figure 6.2.

$$\begin{aligned} \psi_1 &= A_1 e^{\iota k'x} + A_2 e^{-\iota k'x} \\ \psi_2 &= B_1 e^{\iota kx} + B_2 e^{-\iota kx} \\ \psi_3 &= C_1 e^{\iota k'x} + C_2 e^{-\iota k'x} \end{aligned} \tag{6.15}$$

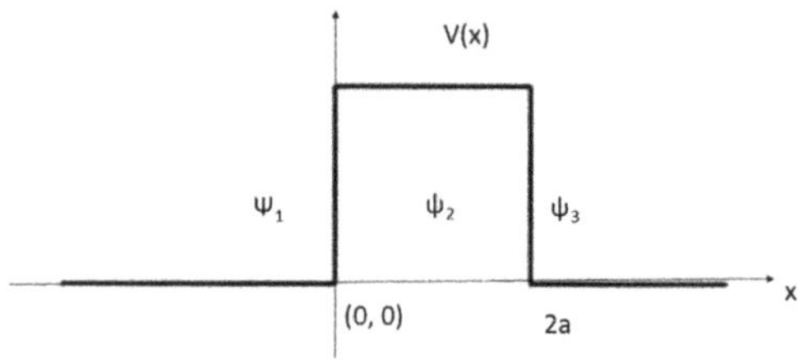

Figure 6.2. Potential barrier for a constant positive potential of width $2a$.

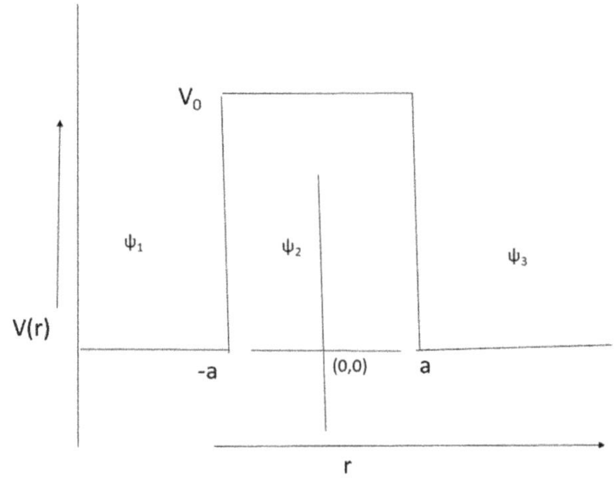

Figure 6.3. Another example of the potential barrier for a constant positive potential of width $2a$.

with

$$k^2 = \frac{2m(E - V_0)}{\hbar^2}, \text{ and}, k'^2 = \frac{2mE}{\hbar^2} \tag{6.16}$$

for $E < V_0$. The corresponding boundary conditions for figure (6.2) is given as:

$$\begin{aligned} \psi_1(0) &= \psi_2(0) \\ \psi_2(2a) &= \psi_3(2a) \\ (\psi')_1(0) &= \psi'_2(0) \\ \psi'_2(2a) &= \psi'_3(a) \end{aligned}$$

Another example of barrier potential is given in figure (6.3) and the related boundary conditions will change to:

$$\begin{aligned} \psi_1(-a) &= \psi_2(-a) \\ \psi_2(a) &= \psi_3(a) \\ \psi'_1(-a) &= \psi'_2(-a) \\ \psi'_2(a) &= \psi'(a) \end{aligned} \tag{6.17}$$

We can then write a general form of boundary conditions for $E < V_0$ (or V_0) in the barrier as:

$$\begin{aligned}\psi_1 &= A_1 e^{\iota k' x} \\ \psi_2 &= B_1 e^{-kx} + B_2 e^{kx} \\ \psi_3 &= C_2 e^{-\iota k' x}\end{aligned} \tag{6.18}$$

for $\psi \to 0$ when $x \to \pm\infty$. The two possible solutions of the above equations are:

$$\begin{aligned}(ka)\tan(ka) &= (k'a) \\ (ka)^2 + (k'a)^2 &= \left(\frac{2mV_0}{\hbar^2}\right)\end{aligned} \tag{6.19}$$

The required solution may be formed for the relevant potential in a suitable coordinate system. Another example of potential barrier is given in figure 6.3. All the results of calculations depend on the choice of origin of coordinate or location of the potential. We cannot cover all the possible examples of potential wells or barriers, but it is important to know that the boundary conditions are changed for the location of these potentials. The comparison of the magnitude of total energy E with that of the potential V help to determine the waveform inside a potential. We just give an example to demonstrate the scheme of calculations.

A barrier with infinite potential indicates a stable bound state that cannot be easily accessed by the external particles unless they have high enough energy to cross the potential barrier created by the binding energy or ionization potential. Classical systems have infinite barriers but quantum mechanics allow tunneling due to finite potential walls. Infinite potential barrier does not allow particles to enter the barrier and the width of the barrier does not matter. So, the infinite barrier can be described as an infinite potential step, and though there is nothing on the other side it creates a step that cannot be overcome, unlike for barriers with limited boundaries.

6.3.3 Step potential

The calculation of probability of reflection and transmission is possible using the step potential as shown in figure 6.4. It is worthwhile to know that the simple case of the one-dimensional Schrödinger equation to infinite potentials (wells or barriers) is not particularly interesting physically. However, it gives the scheme of calculations

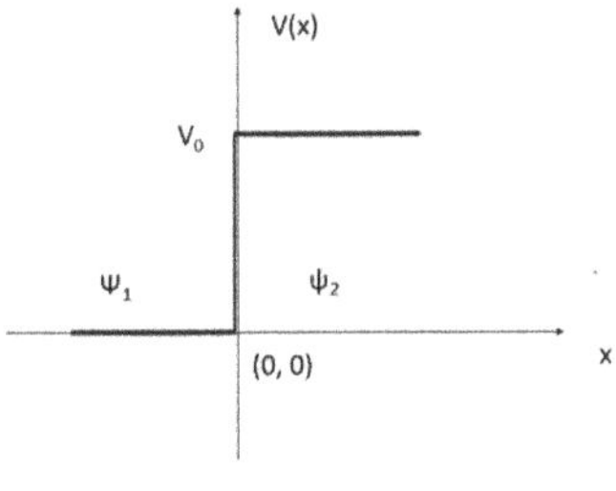

Figure 6.4. Step potential for a constant positive potential of width $2a$.

applied to physical systems. Actual physical systems are much more complicated and all the physical constraints have to be applied. However, all the calculations are needed to understand the behavior of an individual system. A few important steps for extending these calculations to a physical system are:

- Finite potential is compared with the energy of the system. It has to be noted whether the energy is large or small. Classically, energy of the system is always small or large enough to either ignore E or V. A comparison of the energy with potential can let us decide if it qualifies to apply a limit of $V \to 0$ or $V \to \infty$.
- Depending on the shape of the system, the three-dimensional Schrödinger equation is solved. For this purpose, we can choose ∇^2 from a relevant coordinate system, as discussed in chapter 1. The calculations become much more complicated in various coordinate systems. For irregular shapes, the closest shape can be used for approximate results.
- Choice of the correct B.C. is extremely important to understand the appropriate behavior of a system. They depend on the type of potential and its value compared to its energy.
- The choice of B.C. also depends on the origin of the coordinate system and can have a direct impact on the complexity of calculations.

6.3.4 Harmonic oscillator

A general form of the one-dimensional Schrödinger equation for a harmonic oscillator uses Hooke's law in the Lagrangian such that:

$$L = \frac{p^2}{2m} - 1/2m\omega^2(x - x_0)^2$$

such that the Schrödinger equation can be written as:

$$\left[-\frac{\hbar^2}{2m}\frac{\partial^2}{\partial x^2} + \frac{1}{2}m\omega^2(x - x_0)^2\right]\psi = E\psi \equiv \frac{1}{2m}[\hat{p}^2 + (m\omega\hat{x})^2]\psi \tag{6.20}$$

At this point with a little bit of algebra, we can express the Hamiltonian in terms of the raising ($\hat{a}^\dagger \equiv \hat{a}_+$) and lowering operator($a \equiv a_-$) such that:

$$\hat{a}_\pm = \frac{1}{\sqrt{2m\hbar\omega}}(\mp\iota\hat{p} + m\omega x) \tag{6.21}$$

while these operators satisfy the relations:

$$\begin{aligned}
\hat{a}_-\hat{a}_+ &= \frac{1}{2\hbar m\omega}[p^2 = (m\omega x)^2 - \iota(\hat{x}\hat{p} - \hat{p}\hat{x})] \\
\hat{a}_-\hat{a}_+ &= \frac{1}{2\hbar m\omega}[\hat{p}^2 = (m\omega\hat{x})^2 + 1/2] \\
\hbar\omega(\hat{a}_\pm\hat{a}_\mp &\pm \frac{1}{2})\psi = E\psi \\
&\equiv \hat{H}\psi = E\psi
\end{aligned} \tag{6.22}$$

The last equation provides an alternate way to define the various quantum states which defines a second quantization, and allows us to use the Schrödinger equation for many-particle systems, providing a gateway to quantum field theory as discussed later.

6.3.5 Approximation methods in quantum mechanics

Many physical systems have Hamiltonians that cannot be reduced to an exactly solvable part plus a small correction using regular mathematical techniques. In such situations we may use the **variational method** to develop a perturbative series which increases with slowly varying energy. The first term in this Hamiltonian H_0 corresponds to the unperturbed energy which remains there with the system. Perturbation theory in quantum mechanics provides an important tool to solve such problems. We can write the perturbation Hamiltonian which is much smaller than the actual potential, such that the total Hamiltonian of the system H is written as a summation of unperturbed and perturbed Hamiltonian.

The Hamiltonian of a system is written in terms of:

$$H = H_0 + H_p.$$

This expression of H helps to expand the probability of an effect of energy as a function of energy or Hamiltonian operator. The series expansion will only be helpful to solve physics problems if the expansion parameter H_p is sufficiently smaller than H_0 to be able to ignore higher-order contributions. The perturbative part of the Hamiltonian can then be written as $H_p \equiv \sum_{i=1}^{n} \lambda_i W^i$ which provides the perturbative energy which is affecting the system. This perturbation grows with time and the system is being perturbed repeatedly such that the wavefunction keeps on changing and the application of this Hamiltonian on $|\psi(x)\rangle$ is represented as:

$$\psi_n = \phi_n + \lambda^1\phi_n^1 + \lambda^2\phi_n^2 + \lambda^3\phi_n^3 + \cdots$$

giving:

$$E_n = E_n^0 + \lambda^1 E_n^1 + \lambda^2 E_n^2 + \lambda^3 E_n^3 + \cdots$$

$$[(H_0 + \lambda H_p)]|\psi\rangle = [E_0|\psi\rangle + E_n|\psi\rangle_n] \tag{6.23}$$

such that:

$$\phi_n = E_n\psi_n.$$

This perturbation can continue to apply so E_n can be evaluated accordingly. We can write:

$$H_p|\psi\rangle = \sum_1^n \lambda^n W^n|\psi\rangle = E_0\psi + E_1\phi_1 + E_2\phi_2 + \cdots + E_n\phi_n.$$

In the most general form, we can express it as a perturbative series as:

$$|\psi_n\rangle = |\phi_n\rangle + \sum_{m\neq n} \frac{\langle\phi_m|\hat{H}_p|\phi_m\rangle}{E_n^0 - E_m^0} \tag{6.24}$$

and:

$$E_n = E_n^0 + \langle\phi_n|\hat{H}_p|\phi_n\rangle \tag{6.25}$$

The perturbative Hamiltonian can have a constant value identified as a time-independent perturbation theory. There could be a small perturbative potential H_p which can perturb the system and this perturbation will affect the system every time it is perturbed; for example, if an oscillating system is perturbed after an oscillation and continues to oscillate and accelerate or decelerate after each rotation. This perturbation will keep growing even if the perturbation is independent of time. So the wavefunction $|\phi_n\rangle$ will grow into the wavefunction $|\psi_n\rangle$ after n successive perturbations.

Perturbative expansion is one of the most effective approaches to find solutions to quantum mechanical problems. And for sufficiently small perturbation, we can just include the first-order perturbation and ignore all the higher-order effects because of negligible contribution, giving:

$$|\psi_n\rangle = |\phi_n\rangle + \langle\phi_n|\hat{H}_p|\phi_n\rangle$$

All the higher-order terms from equation (6.23) can be ignored in this case. If the perturbation is not that small, a second-order term can also be included. However, the smaller it is the more effective this method of perturbative approach is. For large perturbation we will have to include additional potential in the unperturbed Hamiltonian as $H = H_0 + H_I$, and variational principle or perturbation theory cannot be used. However, a small variation in potential can be treated with time-dependent perturbation theory where potential is a function of time and its variation with time are evaluated to find the energy of the system.

Perturbation theory can be used to solve the equation of motion (differential equation) of a slowly-varying dynamical system. The **Wentzel–Kramers–Brillouin (WKB) approximation** is a mathematical approach to find an approximate solution to linear differential equations with spatially-varying coefficients. It is a semi-classical solution of the Schrödinger equation in quantum mechanics. The approximate form of the wavefunction is an exponential function which can be expanded around small values of semi-classical expansion considering a slowly-varying amplitude or the phase of the wave. So the perturbative expansion of equation (6.23) is actually useful for small perturbations and the variational method is particularly useful in estimating the energy eigenvalues of the ground state and the first few excited states of a system. WKB approximation is applicable for slowly-varying potentials and is sometimes used in solving equations in classical plasma.

Quantum mechanics is applied to tiny objects and their dynamics are usually controlled by electromagnetic interactions. **Tensors** and their transformations have been discussed in the first chapter. When we use tensors for a physical system, we can relate their transformation with the dynamics of the system and express their transformation in terms of potential because the dynamics of the system is controlled by the applied force which is monitored by the change in energy.

It is easy to understand that electromagnetism is much more relevant than gravity at quantum scale due to the extremely small mass of tiny objects and the dominance of electromagnetic interaction over gravity at short distances. However, the interaction between charges, identified as electromagnetic interaction, is spherical in nature. So we have to study the dynamics of an electromagnetically-interacting system in spherical coordinates. One of the required tools to study dynamics of quantum mechanical systems is provided by using three-dimensional tensors in spherical polar coordinates. However, the components of a transformation matrix of tensors in spherical coordinates will depend on the direction cosines of spherical coordinates. The physical interpretation of transformation of spherical tensors is the angular momentum in a spherically symmetric environment.

The Wigner–Eckart theorem then states that the transformation of a tensor operator $\hat{T}_q^k$ in spherical coordinates is transformed using the quantized value of angular momentum, written as:

$$\langle j', m_j' \rangle | \hat{T}_q^{(k)} | j, m_j \rangle = \langle j, k; m_j, q | j', m' \rangle \langle j' || \hat{T}^{(k)} || j \rangle \tag{6.26}$$

where j, m are angular momentum eigenstates and the factor $\langle j', m'_j \rangle | \hat{T}_q^{(k)} | j, m_j \rangle$ called the reduced matrix element depends on the numbers j', m and k.

The simplest application of the Wigner–Eckart theorem is seen for scalars where $k = 0 = q$. Equation (6.26) then reduces to:

$$\langle j', m_j' | \hat{B} | j, m_j \rangle = \langle j = 0, m_j = 0 | j', m' \rangle \langle j" || \hat{B} || j \rangle = \langle j' || \hat{B} || j \rangle \delta_{jj'} \delta_{mm'}$$

The most general solution of Schrödinger equations leads to the quantization of angular momentum which can be used to describe dynamics of electromagnetically-interacting systems with rotation of charge in the field of another charge.

In the case of vector operators we can look at a tensor of rank 1: $T^{(1)} = A^{(1)} = \hat{\mathbf{A}}$ with $A_0^{(1)} = A_0 = A_z$ and $A_{\pm 1}^{(1)} = A_{\pm 1} = \mp(\hat{A}_x \pm \hat{A}_y)/\sqrt{2}$ Inserting these values in cquation (6.26), wc obtain:

$$\langle j', m_j' | \hat{A}_q | j, m_j \rangle = \langle j, 1; m_j, q | j', m_j' \rangle \langle j' || \hat{\mathbf{A}} | j \rangle \tag{6.27}$$

Operator A can be simply replaced by operator J. The quantum mechanical approximations are not limited to time-independent perturbations which can be solved using the variational principle and WKB approximation.

Time-dependent perturbation theory deals with the perturbations and changes with time. In this situation, the perturbative potential may be a function of time as well. However, additional constraints can be applied such that we can study adiabatic approximation or sudden approximation. The transition of states and interaction of radiation with matter can also be treated as time-dependent perturbation theory and most graduate-level quantum mechanics books cover this topic. A detailed description of these topics and the relevant mathematical approach is out of the scope for this book.

6.4 Decay rates and scattering theory

Quantum mechanics describes everything in the form of probability. This probabilistic nature is incorporated into the calculation of decay rates by incorporating the probability of finding the correct states of incident particles and the decay products or scattered particles. Therefore, the decay rates and scattering cross-sections become probability distributions.

6.4.1 Decay rates

We can define the decay rate in terms of the original number of particles N_0 and the decay width $d\Gamma$ as the change in number of particles N per unit time t. Each particle has a characteristic value of the decay time τ. The decay rate τ is defined as $\tau = \frac{\hbar}{\Gamma}$ and corresponds to how long a particle can survive. The rate of change in number of particles depends on the original concentration and the nature of particles such that:

$$\frac{dN}{dt} = -\frac{N_0}{\tau} = \frac{\Gamma}{\hbar}N$$

which has a solution as:

$$N(t) = N_0 \exp\left(\frac{-\Gamma t}{\hbar}\right).$$

This relation is used in any decay process such as low energy nuclear physics or radioactive decays. We can define the decay width $\Gamma(p)$ of a particle and write it as the probability of decay of X particle into two particles A and B such that $X \rightarrow A + B$:

$$\Gamma(X \rightarrow A + B) = \frac{(2\pi)}{2E_i}\int \Pi_{n=i}^{N_b}\frac{dp_n^3}{2E_n}|M_{fi}|^2\delta^4(p_f - p_i) \tag{6.28}$$

This decay width in terms of the probability decides if particle X will decay into $(A + B)$, $(A + C)$, $(B + C)$ or any other combination of particles. The sum of the probabilities of all possible decay modes should be normalized to one to get the relative probability of various decay modes. It can include the two-body decay rate, or three-body decay rate, etc as well. The ratio of one decay mode with the sum of all decay rates is defined as **branching ratio** (BR), such that:

$$BR(X \rightarrow A + B) = \frac{\Gamma(X \rightarrow A + B)}{\Gamma(X)} \tag{6.29}$$

where $\Gamma(X)$ is the total decay rate of particle X which is the sum of all decay rates from all modes.

6.4.2 Scattering cross-section

Scattering is a classical concept which is almost similar to two-body collision. Its quantum mechanical version in terms of point particles is discussed as scattering in three-dimensional space. Scattering theory of quantum mechanics uses the Schrödinger equation to discover the type of interaction in a lot of processes and unveil the structure of targets in scattering processes knowing the incident particles. The elastic scattering obeys a complete set of conservation rules based on the nature of interacting particles. Unlike classical mechanics, quantum mechanical scattering does not conserve the energy and momentum of particles before and after collision. Elastic nature is identified if the incident and final states remain unchanged. Elastic scattering occurs between the same initial and final states, whereas inelastic scattering may occur among different particle states and initial and final states can vary just keeping the required symmetries of the interaction and satisfying the conservation rules among quantum numbers of initial and final states. However, energy-mass conversion can take place using Einstein's famous relation of $E = mc^2$ and invariance of quantum numbers is related to symmetries of the Lagrangian corresponding to the interaction potential or the invariance of the Lagrangian under certain transformations. Conservation rules in quantum mechanics are related to the invariance of the corresponding interaction Lagrangian and its symmetries that define the conservation rules of that Lagrangian.

Before moving further, it is important to understand that quantum mechanics always deals with probabilities. Therefore, all the calculations have a built-in probabilistic approach. The scattering cross-section and decay rates correspond to the probability of scattering or decay, respectively. Therefore, it can deal with the probability of staying inside the potential well or crossing through the barrier. That supports the fact that part of the light reflects and a part is transmitted. All of the above calculations calculate the probability of finding particles in certain states from the modulus square of the wavefunction and integrating it over the corresponding region.

The most interesting scattering is the inelastic scattering which can be visualized as something similar to classical impact physics. However, in quantum mechanics, this is related to the structure of the target and is used to understand the behavior and composition of unknown targets. It is something similar to classical impact physics, but due to quantum mechanical behavior, it gives the information about the compact targets. Depending on the direction of incoming particles, the outgoing particles will move in the opposite direction. However, they can scatter with different angles in the plane perpendicular to the direction of incoming particles and are hence called cross-sections.

The scattering cross-section σ depends on the flux of incident particles and the number of particles in the final states. The small area of cross-section in the perpendicular plane is $\frac{d\sigma}{d\Omega}$, named the **differential cross-section**. The total cross-section of scattering can be obtained by integrating the differential cross-section over the solid angle $d\Omega = d(\cos\theta)d\phi$. The total scattering cross-section is then given as:

$$\sigma = \int \frac{d\sigma}{d\Omega} d\Omega = d(\cos\theta)d\phi.$$

We will discuss the scattering cross-section and decay rates after introducing the relativistic quantum mechanics in later chapters.

Calculation of the scattering cross-section provides mathematical tools to find the structure of the target also. To study scattering theory we look at two-body problem such that the Schrödinger equation for two spinless particles with mass m_1 and m_2 is written as:

$$-\frac{\hbar^2}{2m_1}\vec{\nabla}_1^2 - \frac{\hbar^2}{2m_2}\vec{\nabla}_2^2 + V(\vec{r_1}, \vec{r_2})\Psi(\vec{r_1}, \vec{r_2}) = E_T\Psi(\vec{r_1}, \vec{r_2})$$

where $V(r_1, r_2)$ is the potential and E_T is the total energy. If μ is the reduced mass $\mu = \frac{m_1 m_2}{(m_1 + m_2)}$ of two particles, we can write it as:

$$\left(-\frac{\hbar^2}{2\mu}\vec{\nabla}^2 + V(\vec{r})\right)\Psi(\vec{r}) = E\Psi(\vec{r}) \tag{6.30}$$

The incident wave has potential zero $V = 0$ and satisfies the equation:

$$(\nabla^2 + k_0^2)\phi_{\text{inc}}(\vec{r}) = 0 \tag{6.31}$$

where $k_0^2 = \frac{2\mu E}{\hbar^2}$, with a solution:

$$\phi_{\text{inc}}(\vec{r}) = e^{ik_0\vec{r}} \tag{6.32}$$

whereas the scattered wave interacts with nonzero potential.

6.4.3 Form factors

A form factor is a function in quantum mechanics which encapsulates the particular properties of the interacting particles without going into details of other properties. This function is used to distinguish between a single particle as a point particle and a composite system of particles. Form factor of a point particle is 1, whereas the form factor of any other particle is less than 1. Form factors are identified as structure functions as well. In contrast, the scattered wave has a solution:

$$\phi_{\text{sc}}(\vec{r}) = Af(\theta, \phi)e^{ik_0\vec{r}} \tag{6.33}$$

where $f(\theta, \phi)$ gives the structure of the target or form factor which can change the scattered wave to deviate from plain wave behavior. The total wavefunction of a scattering wave can be expressed as a linear combination of ϕ_{inc} ϕ_{sc} such that:

$$\Psi(\vec{r}) = \phi_{\text{inc}} + \phi_{\text{sc}}$$

Form factors give the actual description of scattering as how incoming states are changed into the outgoing states giving the probability of scattering or the scattering cross-sections in the end.

6.4.4 Born approximation

One of the most popular approaches to determine the scattering cross-section is the Born approximation. When incident energies are sufficiently high, probabilities for all transitions, except elastic scattering, are much less than unity. If, in addition, the initial state is not much changed during the collision, the Born approximation, which employs unperturbed initial and final states, applies. The exact transition amplitude for the transition from eigenstate i to eigenstate n is given by $T = \langle \phi_n | V_f | \psi_i \rangle$ where ψ_i is the exact initial state for the combined system of target and incident particles. The Born approximation gives the scattering cross-section as a Fourier transform of the corresponding potential as:

$$f(\theta) = \int d^3r' e^{-i(k-k')} V(r') \tag{6.34}$$

Apart from this quantitative disagreement, the Born approximation fails to give an exact result as it does not satisfy the optical theorem because it gives the real contribution only and does not satisfy:

$$Im f(0) = \frac{k}{4\pi}\sigma_{\text{total}} \tag{6.35}$$

where:

$$f(0) = \int d^3r' V(r') \tag{6.36}$$

However, $f(\theta)$ makes it possible to generalize this approach for all fundamental interactions.

6.4.5 Distorted wave theories

To include both the Rutherford phase and continuum capture, one may employ continuum states of the projectile rather than the target in the final state. As first discussed by Briggs in the context of electron capture by highly charged ions, a consistent theory emerges by considering expansions of the full amplitude in powers of the small parameter ZT/ZP. This theory has been developed over the year and is known as the distorted wave strong potential Born (DSPB).

6.4.6 Coupled channel approximation

As pointed out in the previous sections, perturbative approaches, like the Born approximation, provide an accurate enough description of ionization processes in fast ion–atom collisions. However, even the Born calculations considerably overestimate total ionization cross-sections, and the difference between calculations and experimental data sharply increases with decreasing collision energy.

Due to their small masses, individual particles can easily acquire relativistic energies. The relativistic effects are incorporated in quantum theory and are extended by quantum field theory that replaces the first quantization by the second quantization that leads to quantization of fields instead of variables. The states are then described by the associated fields instead of the variables of mechanics.

Therefore, at relativistic energies, particle states are attributed by fields instead of variables. However, quantum field theory opens new venues in physics, and is a more effective approach to study all interactions as gauge theories, using the Lagrangian formalism.

We therefore give an overview of all the relevant theories to identify their scope and then link together all the apparently different approaches in the form of QED that becomes a standard gauge theory and provides a framework to study all of the fundamental interactions as gauge theories with their inherent properties at the individual particle level.

Generally, in scattering theory, and in particular in quantum mechanics, the Born approximation consists of taking the incident field in place of the total field as the driving field at each point in the scatterer. The Born approximation is named after Max Born who proposed this approximation in the early days of quantum theory development. It is the perturbation method applied to scattering by an extended body. It is accurate if the scattered field is small compared to the incident field on the scatterer. For example, the scattering of radio waves by a light styrofoam column can be approximated by assuming that each part of the plastic is polarized by the same electric field that would be present at that point without the column, and then calculating the scattering as a radiation integral over that polarization distribution.

6.5 Lagrangian formalism and fundamental interactions

V can be any kind of potential and the theory can be applied to any type of interaction potential. All of the quantum mechanical theories with a microscopic potential can be accommodated in this theory. The action can be calculated as:

$$S = \int L dt.$$

Total action has to be conserved like in classical physics such that:

$$\delta L = 0.$$

This principle is a basis for the variational principle. The equation of motion can be solved for every interaction potential for the relevant boundary conditions and the suitable coordinate system. Writing an equation of motion is the most important step to understanding the dynamics of a system.

Non-relativistic quantum mechanics works for regular subatomic physics and nuclear physics. However, the individual particle processes take place at extremely high energies, and relativistic effects have to be incorporated. However, before getting into the relativistic quantum mechanics, we need to review the basic concepts of relativity. Among the four fundamental interactions, namely gravity, electromagnetism, and the weak and strong interactions, electromagnetism is the most well-understood theory both at the microscopic and macroscopic level. Gravity is not fully understood at the microscopic level, whereas the weak and strong interactions are extremely short-ranged and are not realized outside the nuclear size. Electrodynamics can then be compared with gravity at large scale. Gravity deals with mechanics of all matter and cannot be ignored in the study of dynamics of any material, whereas classical electrodynamics is the study of dynamics of charges and describes the interaction among currents and magnetic fields. Current is basically the rate of flow of charge and has an associated electric and magnetic field. Quantum mechanics is a technique to study the mechanics of tiny objects that simultaneously exhibit the properties of waves and particles.

Atomic and molecular structures can only be studied quantum mechanically. The structure of atoms was the first successful application of quantum mechanics. The interaction of atoms, formation of molecules, and configuration of different microscopic structures of atoms and molecules can be understood very well by quantum mechanics. A large portion of subatomic physics as bound states of electrons and nuclei is well described by quantum mechanics. However, when the dynamics of constituents of atoms like electrons, protons or neutrons are studied as individual particles, their relativistic motion cannot be ignored as light objects. Electrons are especially the lightest form of charged matter and are usually moving with relativistic energies. So relativistic electromagnetism is developed by incorporating relativity into quantum mechanics.

Quantum mechanical study is based on fundamental interactions and discovering the symmetries and conservation rules which are hidden in the potential of interactions. Invariance of a Lagrangian under certain transformations of potential is used to determine the conservation rules of an interaction, which are then associated with the relevant symmetries. This invariance technique also leads to exploring the need to find new symmetries for the identification of the scope of an interaction, which leads to introduction of new symmetries (or quantum numbers). However, these symmetries may not be relevant for other interactions, and the Lagrangian formalism is derived from the constant action principle using the variational principle.

Electromagnetism works at the individual particle level regardless of the scale of a system. It is realized through the charge and the distance between charged objects at any scale. At the microscopic level, however, the quantum mechanical approach is needed to correctly describe it. This generalization of the electromagnetic theory is identified as QED and can be used as the most general theory which can be treated as scale-independent fundamental interaction theory. Among the other three fundamental interactions, we have not yet succeeded in properly generalizing gravity to quantum scale, whereas the weak and strong interactions are not defined beyond the nuclear scale.

Combining quantum mechanics with the quantization of fields (described in quantum field theories), QED emerges as a local theory where each and every field is a function of the four-vector x_μ, which depends on both space and time coordinates and can be transformed into the energy–momentum coordinate system as well. Mathematically, these transformations are represented in terms of unitary groups which identify the nature of the corresponding interaction in terms of its symmetries and the corresponding conservation rules. However, QED has the ability to completely reproduce classical electrodynamics in the classical limit of the Planck's constant tending to zero, which makes it possible to simultaneously measure the position and momentum together or find energy and momentum of a physical system. Moreover, it converts the quantized variables of quantum mechanics and quantum field theory into the corresponding continuous variables of classical mechanics.

Chapter 7

High energy physics and relativity

7.1 Brief overview of particle physics

Fundamental particles are the particles which cannot be divided further at any energy. They are the basic building blocks of matter identified as point particles. They are totally invisible and they are identified by their distinct properties and behavior. Interaction between particles is associated with the intrinsic properties of particles that are called quantum numbers and are defined mainly in quantum field theory at high energy. Most of the quantum numbers describe intrinsic particle behaviors and symmetries lead to the conservation rules. There are only a few fundamental particles, interacting among themselves by four fundamental interactions that are mediated by at least twelve known particles at the subatomic level (or quantum scale). We need to understand fundamental interactions and classify particles according to their properties and their participation in various interactions.

It is worthwhile to mention that all the fundamental particles are fermions and combine together to make atoms. Atomic nuclei are made up of protons and neutrons (called nucleons) and quarks make nucleons. Interaction between quarks is always a nuclear interaction, whereas electromagnetic interaction binds electrons with nuclei to make atoms at quantum scale. Individual properties of particles are hidden in atoms just as individual atoms may not be recognized by their appearance in the molecular form. However, at high energies, individual particles are found independently and their reactions are responsible for the interchange of energy and mass.

All of the fundamental particles are recognized by the ability to realize a particular interaction. Table 7.1 lists the participation of a group of particles in the corresponding interactions. They are identified by their intrinsic properties. The major identification is based on their spin statistics and is used to classify particles in two major groups: fermions (with half-integral spin) and bosons (with integral spin). There are altogether 12 fundamental fermions and 12 known fundamental vector bosons (spin 1 particles). There are six light mass fermions (or leptons) and there are six heavy fundamental fermions (or quarks). Leptons have an associated lepton

doi:10.1088/978-0-7503-6054-8ch7

Table 7.1. Properties of fundamental particles including the mediators. Their charges are given in units of electron charge e, spin in units of $\hbar$, with the abbreviation: QED = quantum electrodynamics, EWI = electroweak interaction, S.I = strong interaction and grav. = gravity.

Particles	Charge	mass	Type	Spin	Interaction
e, μ, τ	$-e$	m_e, m_μ, m_τ	leptons	1/2	em, weak
ν_e, ν_μ, ν_τ	0	0	leptons	1/2	weak
quarks(u, c, t)	+2/3e	m_u, m_c, m_t	baryons	1/2	em, strong
quarks(d, s, b)	−1/3e	m_d, m_s, m_b	baryons	1/2	em, strong
Photon(γ)	0	0	mediate QED	1	em
$W^{\pm}$, Z^0	$\pm 1, 0$	m_W, m_Z	mediate EWI	1	electroweak
8 gluons	0	0	mediate S.I	1	strong
graviton ????	0	unknown	mediate grav.	2 ??	gravity

flavor called lepton number which corresponds to each charged lepton (electron e, muon μ and tau τ) and their corresponding neutrinos (ν) exist in all three different flavors. Heavy particles (or hadrons) have their own six different flavors and are called quarks. These flavors are called down, up, strange, charm, bottom and top represented as (d, u, s, c, b, and t, respectively). All of these along with some other defining properties are listed in table 7.1. All the fundamental particles are bound through sharing their intrinsic properties using conservation rules and the related quantum numbers. Therefore, the dynamics of fundamental particles cannot be understood without a detailed understanding of fundamental forces.

7.1.1 Fundamental interactions and fundamental particles

Dynamics of these particles are defined by fundamental forces which are associated with intrinsic properties of the particles. These forces are always associated with fundamental particles and perform no net work. All of the fundamental forces are conservative in nature and the net amount of work done by these forces is related to the intrinsic properties of matter and is expressed in terms of quantum numbers at the subatomic level except gravity. None of these forces are a contact force and they cannot be turned off in any case. The electromagnetic force is a unique force that is well-understood at the subatomic quantum scale as well as at large macroscopic scale. This is the only tested fundamental interaction at the macroscopic level that has a very well-understood quantum-scale description and its relativistic quantum field theory is known as quantum electrodynamics (QED), which is the most well-developed quantum field theory as well.

The universe is bound together through four fundamental interactions that lead to the formation of large structures. Their shapes and sizes are determined by balancing among multiple forces that have different natures, but their impact may be similar. All of the fundamental forces are conservative in nature and the net work done by these conservative forces always vanishes. They are expressed in terms of intrinsic properties of matter and they obey certain conservation rules and their associated quantum numbers at the individual particle level. All the fundamental forces except gravity are well-behaved at individual particle scale and their integrated effect is observed, directly or indirectly, in a laboratory. The masses of particles are simply added together and observed at the macroscopic scale. At subatomic level, interactions are proposed to be mediated through the exchange of specific mediators which take care of certain gauge symmetries. These mediating particles exhibit particle properties and have all the associated quantum numbers that follow the conservation rules of the corresponding interaction. They are identified by charge, mass and all the relevant quantum numbers for the corresponding theory. Mediating particles for interaction theories are named as intermediate vector bosons and may be created and absorbed virtually during an interaction and can interact with other particles via their individual properties.

There are four fundamental forces in nature. Weak, strong and electromagnetic interactions are observed on quantum scale as gauge theories. Weak and strong nuclear forces are only defined at sub-nuclear scale as gauge theories. However, electromagnetism is described successfully both at the classical and quantum scales. It is dealt with as a gauge theory at quantum scale, whereas it can be defined as a classical theory like gravity and is applicable to the macroscopic level successfully. However, gravity is a well-understood theory at the macroscopic level only. It is still not clearly understood as a quantum theory. If we could describe gravity with quantum theory, we will probably need to describe it with the help of a mediator called the graviton. However, if we succeed in developing a quantum theory of gravity, it will bring more particles in the picture and it will not be observed directly in the visible universe.

There are 12 fundamental particles and there are 12 known mediators of three fundamental interactions which are vector particles in themselves. The theory of mediators (the Yukawa theory) is associated with gauge theories and we have gauge theories of three interactions of strong and electroweak theories. Just to understand it, mediators of electromagnetic interaction are electromagnetic waves that are made up of photons and are represented in the particle form as energy quanta or photons or γ, whereas $W^{\pm}$ and Z^0 correspond to intermediate bosons for weak interaction and eight gluons which make a group of 12 particles altogether and mediate all three fundamental interactions at quantum scale. It is difficult to write a gauge theory of weak interaction. So, we have electroweak theory as a gauge theory. Similarly, if gravity can successfully be described as a gauge interaction, it will be realized by all the massive particles and will have an associated mediator as the graviton. This will have to be a spin 2 particle and will be different from any other mediator.

QED uses the quantization of electromagnetic fields at quantum mechanical scale at relativistic energies to describe electromagnetic interacting of particles locally. So, the study of QED and its scope cannot be determined without a detailed understanding of special relativity, quantum mechanics and electromagnetic interaction. The classical concept of fields and their quantization describes the nature of electromagnetic interaction at the individual particle level and explains how the local theory of QED can be integrated at the macroscopic level. Therefore, a brief introduction of the required theories is included in this book for the relevant background and references for further study are included for convenience.

Particles lose energy during bonding and stay in bound states due to the lack of enough energy to move away from each other and exist separately. The bonding between matter particles can always be broken by a stronger force that can compete with the binding energy to break it apart or convert the particles into energy or convert energy into matter following Einstein's equation. All matter is composed of twelve fundamental particles. Gauge interactions are locally defined and mediators communicate through their specific attributes (or quantum numbers) to mediate all specific interactions. Photons mediate electromagnetic interaction. Weak interaction is mediated by $W^{\pm}$ and Z^0 and the strong interaction is mediated by eight gluons. If we succeed in writing the quantum theory of gravity, then the graviton will be its mediator and will carry spin 2.

Classical electrodynamics deals with electromagnetism using electromagnetic fields and calculates the detailed structure of atoms and molecules. It leads to the bonding of electrons using interaction theory of non-relativistic quantum mechanics. Atomic structure can be described using the classical electrodynamics that deals with electromagnetic waves and particles, using Maxwell's equations. Fundamental principles of electromagnetism are then described as electromagnetic waves and lead to the four-dimensional description of electric and magnetic fields. The relativistic quantum mechanics leads to the quantization of electromagnetic fields and then is described as quantum field theory (QFT) of electromagnetism and is identified as a short name QED.

Gravity and electromagnetic forces are two macroscopic interactions and can be tested directly in laboratories, whereas the weak and strong forces are sub-nuclear range forces and contribute in the formations of nucleons (protons and neutrons). Strong interaction is the strongest known force, whereas weak force is the weakest force at even smaller scale. It is interesting to notice that all the fundamental interactions are governed by radial potential and vary with the separation between interacting particles (r). A comparison of strength and variation of forces with distance are tabulated in table 7.2.

The range of all interactions is given in table 7.2. Some of the universal conservation rules like charge and four-momentum are obeyed by all interactions. Some of the symmetries and conservation rules are called universal rules and are true

Table 7.2. The relative strength and range of different interactions is given with respect to gravity (or strong interaction and its variation in magnitude with respect to distance r).

Interaction	Relative Strength	Range (in m)	Scale	Variation
Strong	1	1	Nuclear	r
Weak	1	1	Nuclear	1
Electromagnetic	$10^{35}(10^{-3})$	∞	Entire space	$\frac{1}{r}$
gravity	$1\ (10^{-38})$	∞	Entire space	$\frac{1}{r}$

Table 7.3. Fundamental interactions and participating particles and mediators.

Interaction	Participating particles	Mediator	Violates
Weak	leptons, quarks, Z, $W^{\pm}$	Z, $W^{\pm}$	several
Strong	quarks, gluons	gluons	nothing
Electromagnetic	charges, photon	photons	isospin
Gravity	masses	graviton ??	unknown

for all interactions such as four-momentum (energy–momentum) conservation. Other rules are associated with particular interactions and the study of conservation rules identifies the interaction itself. A few rules are commonly used to distinguish between different interactions and are given in table 7.3.

Each of these forces follow particular symmetries and conservation rules which are mathematically expressed in terms of the invariance of the corresponding Lagrangian under the change of certain parameters or quantum numbers. Strong interaction, being the strongest, respects all known symmetries and conserves all known parameters of the related gauge theory. Weak interaction holds minimum symmetries and violates maximum number of the known quantum numbers associated with the participating particles, making it the weakest among all gauge interactions. Every quantum number, except isospin, is conserved in electromagnetic interaction and it is the second strongest force after the strong nuclear interaction. Since we do not fully know that how to describe gravity as a quantum theory, we do not know about its behavior at quantum scale.

7.2 A brief overview of special relativity

Relativity is originally introduced and developed as a macroscopic theory of fast-moving objects and applied to celestial bodies in space. The relative motion in the cosmos is controlled by gravity. However, the study of the entire universe requires discussing curved space–time and general relativity is incorporated to understand a spherically-bound universe that is studied in the light of gravity, and obeys the inverse square law of force. However, the dynamics of individual particles and their interaction is studied in special relativity, which is described by the relativistic motion of a particle in its own frame of reference that is identified as Lorentz inertial frame that exhibits constant linear motion with respect to other particles with relative velocities. A summary of special relativity with its important results is given as a brief review of the theory so that we can understand how to treat particles at high energies where transformation between mass and energies is possible. We learn from special relativity that we need to develop four-dimensional formalism to study such processes.

Relative motion of linearly moving objects is a classical concept and the relative velocities are simply calculated using the laws of vector addition. Galilean transformation is the conversion of spatial coordinates from one frame to another frame, keeping length and time unchanged. One-dimensional motion in the x-direction in a stationary frame, under Galilean transformation, is described in a linearly-moving (primed) coordinate system moving with a speed u in the x-direction. The one-dimensional Galilean transformation is given as:

$$\begin{aligned} x' &= x - ut \\ y' &= y \\ z' &= z \\ t' &= t \end{aligned} \tag{7.1}$$

Special relativity is a generalization of Galilean transformation and maintains the invariance of events in terms of spatial length and time interval as the fourth dimension. Then a four-coordinate system is introduced and an event defined in terms of four-dimensional space–time coordinates. Special relativity was developed to incorporate the transformation of electromagnetic signals. This development was based on the inertial frames, which were defined to protect all the laws of classical physics and were named as Lorentz frames. It was also found that the speed of light is constant in all inertial frames. The transformation of electromagnetic processes from the rest frame to the fast-moving frames (with the speed u comparable to the speed of light) is only possible if the transformation of time is also incorporated. So the Galilean transformations were generalized to Lorentz transformations as:

$$\begin{aligned} x' &= \gamma(x - ut) \\ y' &= y \\ z' &= z \\ t' &= \gamma(t - \frac{ux}{c}) \end{aligned} \tag{7.2}$$

where $\gamma = \sqrt{(1 - \beta^2)}$ and $\beta = \frac{u}{c}$. An important criterion is to check the validity of transformation that we need to define the conditions where classical results can be reproduced. For this purpose, we insert $u = 0$ in the above transformation to reproduce classical results. However, with the development of space technology, the indirect test of relativity is now possible through its application to space science. Another application of the special theory of relativity was observed in high energy physics.

Basic postulates of relativity include the fact that light is used as a source of information and the effect of relative motion on light is different than for material objects. Light moves much faster than everything else in the universe. Light has the maximum speed and is the main source of information.

- Speed of light remains the same in all frames of reference.
- Laws of physics remain invariant in all frames of reference that exhibit linear motion with respect to each other.

Inertial frames or Lorentz frames are then defined as the frames that keep the laws of physics invariant under Lorentz transformation. Such frames are moving with a uniform linear velocity with respect to each other and the relative speed is always lower than the speed of light. Lorentz frames exhibit linear motion under Lorentz transformations (7.2) and guarantee the invariance of four-dimensional length instead of three-dimensional length. Four-dimensional length can be related to an event which occurs in a given space for a given interval of time and is described as:

$$x'^2 = x^2 = x_1^2 + x_2^2 + x_3^2 - (ct)^2$$
$$= (x'_1)^2 + (x'_2)^2 + (x'_3)^2 - (ct')^2$$

The fourth coordinate is an imaginary coordinate ($x_4 = \iota ct$) and its square is subtracted instead of addition. Four-dimensional length of a four-vector physically corresponds to an event. The invariance of three-dimensional length indicates difference between two points in three-dimensional coordinates and is given as:

$$L^2 = \Delta X^2 = (x_2 - x_1)^2 + (y_2 - y_1)^2 + (z_2 - z_1)^2$$

where x, y and z are the regular spatial coordinates. The three spatial coordinates (x, y, z) are generalized to four-dimensional space with $(x, y, z; \iota ct)$ coordinates in flat space–time system in special relativity. Four-dimensional change in length is a straightforward generalization of the above relation and is given as:

$$L^2 = \Delta X^2 = (x_2 - x_1)^2 + (y_2 - y_1)^2 + (z_2 - z_1)^2 - c^2(t_2 - t_1)^2$$

It is noticed that the relative motion causes the length contraction for an observer in a frame outside the rest frame of the object that is moving with a speed comparable to the speed of light. On the other hand, the time dilation is observed as the event appears to slow down for the relatively moving observer. The length contraction and time dilation between two frames moving with the relative speed u is given as:

$$\Delta x' = \gamma \Delta x \tag{7.3}$$

where the event is taking place in the rest frame represented as unprimed frame S and the observer is in a moving frame S' that is moving with a uniform velocity u with respect to S. Both S and S' frame are inertial frames as they obey the law of inertia, which states that a frame at rest will remain at rest and a moving frame will keep on moving with uniform relativistic speed in the absence of external force. The time span of the event is transformed as:

$$\Delta t' = \frac{\Delta t}{\gamma} \tag{7.4}$$

The correspondence principle between special relativity and classical physics can be easily verified when $u \ll c$, with 'c' the speed of light. For nonzero speed c is conserved for $c = \frac{\Delta x}{\Delta t} = \frac{\Delta x'}{\Delta t'}$ such that the contraction in length is balanced by the dilation of time. Using the above transformations, the invariance of laws of physics can be checked.

The failure of the Galilean transformation was associated with electromagnetism mainly because the laws of electrodynamics could not be proved invariant under Galilean transformation. When simultaneous measurements are possible for both the rest frame and the moving frame, then Galilean transformation works. However, that simultaneity is only true for a single point in space. Otherwise, synchronization of clocks in the rest frame are not necessarily synchronized in the observer frame in a moving frame.

Light, as an electromagnetic wave, moves at a constant speed which depends on the electric permittivity ε_0 and magnetic permeability μ_0 in free space such that:

$$c = \frac{1}{\sqrt{\varepsilon_0 \mu_0}},$$

and the speed of light is accepted as a universal constant as its measurable value in free space is independent of the frame of reference.

The four-dimensional formalism of special relativity is evolved through the Lorentz transformation and describes the invariance of an event in different frames of reference. Galilean transformation kept time constant in all frames of reference as if simultaneous measurement of events was possible in all frames. The Lorentz transformation addresses the time transformation along with the transformation of spatial coordinates such that a compensation between length contraction and time dilation take place to conserve the four-dimensional length. Lorentz transformation between two frames S and S' moving in the x-direction with respect to each other is given as:

$$\begin{aligned} (x'_1 - x'_2) &= \gamma(x_1 - x_2) \\ (y'_2 - y'_1) &= (y_1 - y_2) \\ (z'_2 - z'_1) &= (z_1 - z_2) \\ (t'_2 - t'_1) &= \frac{(t_2 - t_1)}{\gamma} \end{aligned} \tag{7.5}$$

Lorentz transformation shows the space–time invariance of event's location and duration (four-dimensional length). In compact form, these transformations are represented as:

$$\begin{aligned} dt &= \gamma(dt' - udx'/c^2) \\ &dx(dx' + udt') \\ dy &= dy' \\ dz &= dz' \end{aligned} \tag{7.6}$$

The Lorentz invariance can relate length contraction and time dilation to develop a correspondence between relativity and classical mechanics such that the invariance of laws of mechanics in four-dimensional formalism is fully satisfied. However, the large scale application of special relativity is not relevant for slowly moving objects ($u \ll c$) such that the Lorentz invariance cannot be tested due to the simultaneous measurement of different points. In regular daily events or in most labs, the relativistic effects are negligible even if simultaneous measurements are not possible. The Lorentz transformation and postulates of relativity show some interesting results which are not described without using relativity. It includes the law of transformation of velocities that depends on the direction of motion, as expected. Some of the important results of relativity are listed below:

$$\begin{aligned}\frac{dx}{dt} &= \frac{\gamma(dx' + udt')}{\gamma(dt' - vdx'/c^2)}\\ \frac{dy}{dt} &= \frac{dy'}{\gamma(dt' - udx'/c^2)}\\ \frac{dz}{dt} &= \frac{dz'}{\gamma(dt' - udx'/c^2)}\end{aligned} \tag{7.7}$$

These equations lead to the velocity transformation relations for the corresponding components of velocities in a moving frame in the x-direction and are observed from the corresponding rest frame as:

$$\begin{aligned}v_x &= \frac{v'_x + u}{1 + uv'_x/c^2}\\ v_y &= \frac{v'_y \times \gamma}{1 + uv'_x/c^2}\\ v_z &= \frac{v'_z \times \gamma}{1 + uv'_x/c^2}\end{aligned} \tag{7.8}$$

where v is the velocity of the object in the Lorentz frame S and v' is the transformed velocity of the object in the S' frame. Transformation of velocities leads to the change in kinetic energy which appears in the form of change of frequencies and wavelength. The change in frequency and wavelength associated with the change of velocities is called the Doppler effect. This effect states that a decrease in wavelength takes place when objects move towards each other and is called a blue shift, whereas an increase in wavelength, or red shift, can occur when the source and observer are moving away from each other. If u is the velocity of the object carrying light of frequency f, the observed frequency f_{ob3} is given by:

$$f_{\text{obs}} - f\sqrt{\frac{1 - u/c}{1 + u/c}} \tag{7.9}$$

whereas for blue shift the source is moving towards the observer such that:

$$f_{\text{obs}} = f\sqrt{\frac{1 + u/c}{1 - u/c}} \tag{7.10}$$

Until now, relativity is applied to macroscopic level and the concept of Lorentz frames is applicable to non-accelerated frames of reference and this approach is called special relativity. Accelerated frames can be ignored for normal use of physical systems in a laboratory. However, for the study of cosmology, accelerated frames are essential to incorporate curved space–time. General relativity cannot be neglected for celestial bodies. However, we are not including general relativity here as it is only applicable at large scale and we deal with tiny objects at quantum scale.

Special relativity is one of the approaches which makes the study of fast-moving tiny objects more accurate. There is no way to restrict an individual particle motion to a non-relativistic scale. The particle energies in nuclear processes always indicate relativistic velocities. Even in atoms, electrons are moving with sufficiently high energies where relativity cannot be separated from quantum mechanics. Special relativity actually provides a framework to study all the individual particle processes from the rest frame of a particle. This is how the birth of relativistic quantum mechanics led to the development of quantum field theory as a framework of particle interaction theory. A physical interaction theory is developed as a gauge theory and the invariance (symmetries) of the Lagrangian of interaction under certain transformation leads to defining the conservation rules of a theory.

Einstein's theory of special relativity points out that even mass is not a constant of motion for relativistic systems. The mass–energy relation indicates that a mass can convert into energy if it moves with relativistic speed. On the other hand, energy can be converted into mass by absorption of radiation. Relativity states that mass can be converted into energy if it moves with the speed of light, and the energy created by the movement of mass creates the energy such that:

$$E = mc^2 \tag{7.11}$$

Equation (7.7) is the most relevant relation of relativity for individual particles moving with extremely high energies and can be tested in labs as well. Conservation of energy allows the change of the form of energy so the kinetic energy of particles with large velocities can be converted into potential energy and vice versa. Kinetic energy of particles can then contribute to potential energy in the form of mass and kinetic energy of relativistic particles can add mass on slowing down such that the total energy satisfies a relation:

$$E^2 = p^2c^2 + m^2c^4 \tag{7.12}$$

Equation (7.12) assigns two values of energies to particles: positive energies corresponding to incoming states or particles and negative energies corresponding to outgoing states as antiparticles. Most of the other quantum numbers are inverted for antiparticles as well. In a simple way, if an incoming state is added to a system as a particle then an outgoing particle is considered as a subtraction of a particle or an addition of antiparticle. This means that any additional quantum number is subtracted and the momentum of an antiparticle is opposite to the momentum of particle.

This book is comprised of a comprehensive review of electromagnetism, developed in the twentieth century and its role in the development of modern technology. This work is still in progress and understanding of gravity as a quantum theory is the most challenging part. However, several competing theoretical models are in place to understand quantum gravity, but experimental justification still has a long way to go and the existing models are going through improvement in the light of observational data and experimental results. However, before incorporating relativity in quantum mechanics to develop QFT, we need to briefly introduce particle physics as a subject to be able to use various types of matter particles to justify relativistic description of particles at the quantum scale. For this purpose, electrodynamics can be represented as QED.

7.3 Interaction theory

Among all of the fundamental interactions, electrodynamics is the most well-understood theory, both at the microscopic and macroscopic levels. Gravity is not fully understood at nuclear scale, whereas the weak and strong interactions are extremely short-ranged and are not realized outside the femtometre (10^{-15} m) scale. Electrodynamics can then be compared with gravity at a large scale. Gravity deals with mechanics of all matter and is relevant in context with the study of dynamics of matter, whereas classical electrodynamics is only relevant for the dynamics of charges and describes the interaction among moving charges in the form of currents with magnetic fields. Current is basically the rate of flow of charge and has an associated electric and magnetic field. Quantum mechanics is a technique to study the dynamics of tiny objects that exhibit the properties of waves and particles simultaneously.

Electrons are extremely light particles and rapidly revolve in the nucleus as well as spin around their own axes of rotation in clockwise or counterclockwise direction to create a stable system. The spin degrees of freedom add correction terms in the angular momentum, which becomes more complicated with the increase in number of particles. Electronic orbitals deal with individual negative charges in the vicinity of other revolving negatively-charged electrons, incorporating their individual spin and obeying certain selection rules. That is the reason that atomic orbits are much more complicated and cannot be described without quantum mechanics.

Individual particles, due to their tiny masses, can easily acquire relativistic energies and relativity has to be incorporated with quantum mechanics which leads to the quantization of associated fields that can be considered as canonical variables. However, relativity needs to extend the ordinary three-dimensional analysis to four-dimensional canonical variables including space and time in coordinate space. Four-dimensional coordinate space has three spatial coordinates and one time coordinate as:

$$x_\mu = (ict, \overrightarrow{x})$$

whereas the corresponding set of four-dimensional (conjugate) momentum space combines three components of a momentum vector with one component of energy E such that:

$$p_\mu = (iE/c, \overrightarrow{p})$$

The wavefunction of quantum mechanics is now dependent on four coordinates and due to the relativistic motion of particles, the wavefunction is now treated as a field where the quantized form of the field represents a particle in a phase space. Now the four-dimensional parameters work as field operators and corresponding coordinates are expressed as fields. This phenomenon is called **second quantization** or

canonical quantization. The state vectors of quantum mechanics are then replaced by canonical fields instead of the relativistic variables of mechanics.

The fermions vector field ψ in itself has two components (spinors) as $u(p)$ for particles and $v(p)$ for antiparticles. Each one of these spinors has two components as spin-up (+1/2) and spin-down (–1/2). Quantum field theory uses the Lagrangian formalism that gives a description of fundamental interaction in gauge theories by specific potential and symmetries and conservation rules of the theory are derived from the invariance of the Lagrangian.

It is very important to note that particle physics studies the dynamics of fast-moving individual particles in terms of fundamental interactions. For this purpose, quantum field theory is used, which employs relativistic quantum mechanics, and the particles in the theory are described in terms of quantized fields and not space- and time-dependent wavefunctions. Fields are four-dimensional vectors and they are operated by four-dimensional operators to describe the dynamics of particles. A combined theory of quantized field operators and quantized wavefunctions is QFT.

7.4 Relativistic quantum mechanics

A detailed structure of matter and the dynamics of particles at subatomic level has to be studied in terms of quantum mechanics that is primarily based on the wave–particle duality or the uncertainty principle. Quantum mechanics incorporates all possible states of a system with the help of probability at the expense of precision and prefers to know about all possible states. For this reason, particles are represented as state functions, describing the particle and wave properties simultaneously.

On the other hand, the microscopic form of electromagnetic interaction has to coordinate with the same behavior for individually charged particles at relativistic energies and describe composite charges and the associated laws of classical electromagnetism. Individual charges require a quantum mechanical approach for such systems. However, at high energies, simultaneous application of relativity, quantum mechanics and electrodynamics without ignoring the basic principles of either one of these theories develops some new features. Such a complete theory which fully describes the electromagnetic interaction of relativistic systems at quantum scale is called quantum electrodynamics and is a gauge theory and can be treated as QFT of an electromagnetically-interacting system.

Motion of material objects in three-dimensional space takes place with time and it is studied in terms of material properties and energy of the system. Fundamental forces provide energy to do work and transfer energy without particles touching each other. Electromagnetic energy is transferred by radiation. The kinetic energy of motion may be provided by the applied force, which could be mechanical, electromagnetic or any other kind of force. Energy is measured by the changing behavior of matter and it can change its forms as well. Mass can be converted into energy by relativistic motion and energy converts into mass by the loss of energy into mass and slows down, whereas the electromagnetic waves correspond to electromagnetic energy and wave–particle duality proves that it corresponds to a massless particle.

At the macroscopic level, electromagnetism deals with current and charge and obeys certain laws of electromagnetic theory. This form of electromagnetism gives measurable principles in the lab and deals with the macroscopic objects directly. This is called classical electrodynamics that plays a key role in the development of modern technology. This technology includes basic motors, home appliances, digital circuits, supercomputers, electric cars, airplanes and space shuttles, which all work under the principle of electrodynamics in a way.

The development of QED is done for a single particle like any other theory, but the application of the theory needs a generalization to many-particle systems. QED deals with identical particle systems which are identified by different states that can be distinguished, mainly by spin. Therefore, the application of statistical mechanics is naturally justified. Probability theory is inherited in QED from quantum mechanics. Probability theory and statistical analysis is obviously convenient for an ensemble of identical particles residing in different identical states which can accommodate identical particles indistinguishably. However, QED deals with two different classes of particles for relativistic systems such that the integral spin bosons and the half-integral spin fermions obey different spin statistics and have to be described by two entirely different equations of motion named as the Klein–Gordon and Dirac equations, respectively. Both of these equations are derived in the next chapter and discussed in detail.

Relativistic quantum mechanics studies the dynamics of charges that move with relativistic energies. Electrons move with high speed due to light mass and move with high enough speed, comparable to the speed of light. At high energies, interconversion of mass and energy can take place. Relativity requires the study of particle dynamics in four-dimensional space and treats energy and momentum as components of the four-vectors in momentum space and they are related to each other using Einstein's equation as $E = mc^2$. In this case, total energy attains the relativistic form as:

$$E^2 = p^2c^2 + m^2c^4$$

The above equation has two solutions to describe the dynamics of highly-energetic systems with positive and negative energies. The concept of antiparticles is brought in relativistic theories from negative energies. Antiparticles have all the quantum numbers with opposite polarity and correspond to outgoing states as compared to incoming particle states. Condensed matter physics uses the concept of holes for the absence of electrons or vacant states where electrons can reside inside atoms, indicating the available states for incoming electrons, but antiparticles are particles with negative energies and are defined by all the associated quantum numbers. Therefore, relativistic processes involve particles with positive energies and antiparticles with negative energies. This concept of positive and negative energy solutions can only be accommodated at relativistic energies, where we can look at the process from different inertial frames such that an outgoing particle can be treated as a particle with all the inverted quantum numbers and is identified as antiparticle. However, antiparticles are real particles with opposite properties.

Electrons are the lightest known type of matter with a negative charge. These charged particles are highly energetic but have dominant electromagnetic interaction over gravity. Electrons need relativistic contributions in quantum mechanics and ignore gravity. For a proper development of QED, we need to have a good understanding of quantum mechanics, electrodynamics and relativity. However, we first need to know special relativity and Einstein's mass–energy relation to develop four-dimensional formalism in quantum mechanics. There follows a brief review of basic principles of electromagnetism, key concepts of quantum mechanics and basic concepts of relativity to get ready to develop QFT of electrodynamics interaction, named quantum electrodynamics.

7.4.1 Relativity and electrodynamics

Atomic and molecular structures can only be studied quantum mechanically. Structure of atoms was the first successful application of quantum mechanics. Interaction of atoms, formation of molecules, and configuration of different microscopic structures of atoms and molecules can be understood very well by quantum mechanics. A large portion of subatomic physics is bound states of electrons and nuclei and is well described by quantum mechanics. However, when the dynamics of constituents of atoms like electrons, protons or neutrons is studied as individual particles, their relativistic motion cannot be ignored as light objects. Electrons are especially the lightest form of charged matter and are usually moving with relativistic energies. So relativistic electromagnetism is developed by incorporating relativity in quantum mechanics.

Relativistic quantum mechanics adopts the four-dimensional formalism of relativity. However, in this four-dimensional formalism particles are treated as fields and fields are used as variables. The quantization of fields is called second quantization and this form of relativistic quantum mechanics is called QFT. Since relativity and quantum mechanics both reproduce the results of classical mechanics so QFT is the most generalized theory to describe dynamics of material objects and is the perfect theory for all interactions at extremely small scale. This theory gives the most detailed description of fundamental forces at the individual particle level. All the fundamental forces are conservative in nature and QFT is a proper description of all the fundamental interactions except gravity.

QED deals with the study of electromagnetic interaction of individual particles and it is developed as a gauge theory. However, for a detailed understanding of this approach, a brief overview of quantum mechanics and classical electrodynamics is required to develop some special mathematical tools with their physical interpretation to make this theory more effective.

7.4.2 Four-dimensional formalism as a roadway to QED

Maxwell's equations use fundamental laws of electromagnetism, which indicate that light, electricity and magnetism are really a manifestation of the same phenomenon. Maxwell's equations provide a way to write the wave equation for electromagnetic waves. Maxwell's equations provide a way to express electromagnetic field in the form of waves and for the first time introduce electromagnetic waves, opening up a gateway to modern technology. This is where the four-dimensional formalism of electrodynamics starts that leads to the development of QFT.

QED is a combined theory of quantum mechanics and electrodynamics of highly energetic particles moving with relativistic velocities. Classical electrodynamics is a classical description of the dynamics of electromagnetically-interacting charges. It deals with continuous (electric and magnetic) fields that are associated with charge and its motion. Relativity deals with the motions of systems moving with speeds comparable with the speed of light. However, the solution of the equation of motion of non-relativistic quantum mechanics of microscopic objects gives the quantization of angular momentum and other variables due to wave–particle duality, whereas the quantum mechanics of highly energetic particles with relativistic velocities leads to the quantization of fields and is called second quantization.

For a proper understanding of QED, we need to have a good understanding of quantum mechanics, electrodynamics and relativity. However, we only need special relativity and Einstein's mass–energy relation to develop four-dimensional equation mechanics as a relativistic theory of quantized fields. In this chapter, there is a brief overview of the basic principles of electromagnetism, key concepts of quantum mechanics and basic concepts of relativity to get ready to develop QFT quantum field theory of electrodynamics interaction, named QED.

QED uses the quantization of electromagnetic fields at quantum mechanical scale at relativistic energies to describe an electromagnetically-interacting particle. The understanding of quantum electrodynamics and its scope cannot be determined without a detailed understanding of special relativity, quantum mechanics and electromagnetic interaction. The classical concept of field and its quantization describes the nature of electromagnetic interaction at the individual particle level and explains how QED can be integrated at the macroscopic level. Therefore, a brief introduction of all the fundamental theories is included in this book for the relevant background and references for further study are included for readers' convenience.

A combined theory of relativity and quantum mechanics was called relativistic quantum mechanics, which could incorporate spin into quantum mechanics. Discovery of the intrinsic spin of particles led to the classification of particles based on spin. Particles with integral spin are called bosons and with half-integral spin are called fermions. Bosons and fermions were found to follow different spin statistics at high energies, and the Schrödinger equation had to be rewritten in two different equations in relativistic quantum mechanics: namely, the Klein–Gordon equation and the Dirac equation, respectively.

7.4.3 Four-dimensional representation of Maxwell's equations

The vector nature of both fields has to be expressed in terms of a matrix so the electromagnetic field can be fully represented as a matrix in four-dimensional space as $F_{\mu\nu}$ in terms of vector potential:

$$F_{\mu\nu} = \frac{\partial A_\mu}{\partial x^\nu} - \frac{\partial A_\nu}{\partial x^\mu} \tag{7.13}$$

$$F_{\mu\nu} = \frac{1}{c}\begin{pmatrix} 0 & E_1 & E_2 & E_3 \\ -E_1 & 0 & -cB_1 & cB_2 \\ -E_2 & cB_1 & 0 & -cB_3 \\ -E_3 & -cB_2 & cB_3 & 0 \end{pmatrix} \tag{7.14}$$

In Minkowski space, where we take time real and space imaginary, we can represent:

$$\partial^\mu = \left(\frac{\partial}{\partial(ct)}; \frac{\partial}{\partial(x_1)}, \frac{\partial}{\partial(x_2)}, \frac{\partial}{\partial(x_3)}\right)$$

The electromagnetic field tensor is an antisymmetric tensor such that $F_{\mu\nu} = -F_{\nu\mu}$ at relativistic energies, and it leads to a simple set of four Maxwell's equations in a compact form in four-dimensional space as:

$$\partial^\mu F_{\mu\nu} = 0 \tag{7.15}$$

A detailed look at these equations together gives:

$$\partial^\mu F_{\mu\nu} = \partial^\mu\left(\frac{\partial A_\mu}{\partial x^\nu} - \frac{\partial A_\nu}{\partial x^\mu}\right) \equiv \frac{j_\nu}{c} \tag{7.16}$$

such that the four-current conserves in physical systems.

7.4.4 Maxwell's equations in four dimensions

Maxwell's equations are the fundamental equations of electromagnetism and they are used to describe the dynamics of charge at non-relativistic energy. Equations (7.15) and (7.16) are the four-dimensional forms of Maxwell's equations which are applicable to daily life. These equations work at non-relativistic energies. However, we know that change of position potentially incorporates the use of time and we add time as an imaginary component. Similarly, every charge has the ability to create a magnetic field, so we can define a four-dimensional field associated with charge with three real components of electric field $\vec{E}$ and an imaginary component as the magnetic field vector $\vec{B}$ which is always perpendicular to the electric field. The relation between electric and magnetic field is expressed in terms of the Maxwell's equations. Equations (7.13)–(7.16) together give a set of differential forms of the Maxwells'equations and use all the important laws of electromagnetism that give birth to modern technology. Equation (7.15) leads to:

$$\vec{B} = \vec{\nabla} \times \vec{A} \tag{7.17}$$

where A is called a vector potential.

The differential form of **Maxwell's equations**, in free space, then read:

$$\begin{aligned}
\vec{\nabla}\cdot\vec{B} &= 0\\
\vec{\nabla}\cdot\vec{E} &= 4\pi\rho\\
\vec{\nabla}\times\vec{E} &= -\frac{1}{c}\frac{\partial\vec{B}}{\partial t}\\
\vec{\nabla}\times\vec{B} &= \frac{4\pi}{c}\vec{J}+\frac{1}{c}\frac{\partial\vec{E}}{\partial t}
\end{aligned}\tag{7.18}$$

These relations tell us how the variations in the electric and magnetic field depend on each other. The time dependence of the magnetic and electric fields, in reference to Faraday's law and Ampère's circuital law, relate the time-varying electric and magnetic fields to the electric and magnetic flux. Without getting into detailed discussion of Maxwell's equations, we can look into the corresponding integral form of these equations as:

$$\begin{aligned}
\oint_{\partial\Omega} \vec{E}\cdot d\vec{S} &= 4\pi \iiint_{\Omega} \rho dV\\
\oint_{\partial\Omega} \vec{B}\cdot d\vec{S} &= 0\\
\oint_{\partial\Sigma} \vec{E}\cdot d\vec{l} &= -\frac{1}{c}\frac{d}{dt}\iint_{\Sigma} \vec{B}\cdot d\vec{S}\\
\oint_{\partial\Sigma} \vec{B}\cdot d\vec{l} &= \frac{1}{c}\left(4\pi\iint_{\Sigma}\vec{J}\cdot d\vec{S}+\frac{d}{dt}\iint_{\Sigma}\vec{E}\cdot d\vec{S}\right)
\end{aligned}\tag{7.19}$$

Classical electrodynamics works perfectly fine at macroscopic level just as classical mechanics works perfectly fine at the same scale where we put a constraint on the overall behavior of physical systems. Classical mechanics uses Newton's laws of motion for linear systems, Kepler's laws for the dynamics of orbital motion of rotating systems bound by gravity, and fundamental laws of electrodynamics for electromagnetically-interacting systems. Gravity and electromagnetic force are both central forces and create gravitationally-interacting orbits in space and electromagnetically-interacting orbits in atoms. We just consider the dominant force and the angular momenta for the formation of orbits.

The equation of motion of classical mechanics for linear motion is written in terms of the rate of change of linear momentum, whereas the rate of change of angular momentum in circular motion describes the equation of motion for rotation. Net linear force is related to the linear acceleration and net torque on a body is related to the angular momentum. Classical physics deals with continuous variables of a system, in principle, and can attain any value and net force has no restriction on how much acceleration can be produced.

A major difference between gravity and electromagnetism is associated with the basic properties of matter. Mass has only one degree of freedom and gravity is always an attractive force that depends on the quantity of mass and the shape of the objects, whereas electric charges have polarity and could be either positive or negative, which gives two types of behavior to the force, either attractive or repulsive. Charged particles still have mass and the mechanical forces are still there along with the electromagnetic force.

Gravity generates orbital motion, gravitational attraction competing with the centrifugal force generated by the angular momentum. Gravitational orbits have no limit on the amount of mass, it just need a large enough force between two masses to ignore any other external force, even the gravitational pull of other objects. Electrodynamics does not produce orbital motion at macroscopic scale as mass does not carry enough net charge to even move to a measurable distance with large mass. In neutral matter, gravity dominates over electromagnetic force at large distances due to the electrical neutrality of independently existing matter. Neutral matter can only carry partial induced charge, which is not only weak but also is not permanently there. Independently existing neutral matter can be generated by collecting all the positive charges in the center and letting an equal number of negatively-charged light particles revolve around it inside the atoms. Therefore, the electromagnetic forces can only generate orbitals of light electrons in atoms.

Atoms are electrically neutral and are composed of an equal number of electrons and protons. Protons reside with neutrons in the central part of the atom in the nucleus. Neutrality of atoms is indebted to the polarity of charges as an equal number of charges can balance out to give overall neutral matter. However, the polarity of charge makes it more complicated because opposite charges cancel each other and net charge vanishes. Repulsion between similar charges has to be managed in orbital motion of electrons and the repulsion between protons in the nucleus is controlled by strong interaction between protons and neutrons. Without getting into the details of atomic structure, the nucleus can be considered a single positively-charged center and electrons revolve around that center. Almost all the mass of an atom resides in the nucleus and light electrons revolve around it; gravitational interaction is easily ignorable in atoms.

The formation of orbitals of charged particles can occur in the special configuration of atoms only where electrons can revolve around the nucleus but repel one another as well. Moreover, revolving charges exhibit a small magnetic field due to their rotational motion. So the electronic orbits not only keep the balance by matching the electromagnetic force and the angular moment, but they also take care of the magnetic moment of electron due to the orbital motion of electrons. This situation is managed by the quantization of angular momentum in atoms and discrete energy levels are defined although all of the electrons are identical particles and the size of the nucleus is small enough to be treated just as a central point.

Because electrons are light particles that revolve around the nucleus and spin around their own axes of rotation, their spin degrees of freedom add correction terms to the orbital angular momentum that are called spin angular momentum such that the total angular momentum $\overrightarrow{J}$ is a vector sum of orbital angular momentum $\overrightarrow{L}$ and spin angular momentum $\overrightarrow{S}$ which becomes more complicated with the increase in number of particles. Electronic orbitals deal with individual particles so it has to incorporate individual particle properties like spin as well. That is the reason that the electron orbits are much more complicated and cannot be described by some laws similar to Kepler's laws of orbital motions (based on central force and angular momentum).

Due to the light mass of electrons and tiny size of atoms, classical physics is not enough to describe the electronic motion. Quantum mechanics is needed to understand the motion of electrons inside the atoms. The light and tiny electrons move too fast to ignore relativity as well. A detailed study of atomic structure is made possible using quantum mechanics. The key concepts of quantum mechanics are developed using the uncertainty principle in the light of wave–particle duality and the operator formalism is instrumented for this purpose. The probabilistic nature of quantum theory, due to the limitations in precise measurements, does not allow easy discrimination among different possible states of electrons. The equation of motion of electrons in quantum mechanics is the Schrödinger equation and its solution in spherical polar coordinates leads to the discrete value of angular momenta. This quantization of angular momentum and energy is called the first quantization associated with the quantization of state variables. These quantized variables are identified as quantum numbers and the particles are represented as state functions that are described by the particles as well as wave properties, as discussed in the chapter of quantum mechanics.

Light particles such as electrons, due to their small masses, easily acquire relativistic velocities. Non-relativistic studies may not give complete information. Even though the speed of electrons inside atoms is large, non-relativistic quantum mechanics works fine. However, the electrons acquire very high velocities at high energies and even the other particles need relativistic treatment. Therefore, special relativity is incorporated in quantum theory which leads to the second quantization or quantization of fields instead of quantization of variables. Relativistic quantum mechanics evolves into QFT that replaces the first quantization by the second quantization and shows the quantization of fields instead of variables. Therefore, at relativistic energies, particle states are attributed by fields instead of variables. QFT needs to incorporate spin of the particles in its analysis. However, QFT opens new venues in physics and acquires a more effective approach to study interaction as a local gauge theory using Lagrangian formalism.

We therefore give an overview of all the relevant theories to identify their scope and then link together all the apparently different approaches in the form of QED that becomes a standard gauge theory and provides a framework to study all of the fundamental interactions as gauge theories with their inherent properties at the individual particle level. In QFT, the particle spin has to be incorporated and the Schrödinger equation, the equation of motion of quantum mechanics, gives rise to two different equations of motion, namely, the Klein–Gordon equation for particles with integral spin or spin zero and the Dirac equation for half-integral spins. The gauge invariance requirement of the Lagrangian leads to the discovery of several more quantum numbers.

Part III

Development of quantum electrodynamics

Chapter 8

Development of quantum electrodynamics

8.1 Introduction

Quantum electrodynamics (QED) is a theory which incorporates relativity into quantum mechanics to study fundamental interactions at the individual particle level. The generalization of the Schrödinger equation in a four-dimensional relativistic coordinates system leads to the quantization of interacting fields at relativistic energies. The classical form of the total energy E is written as a sum of the kinetic and potential energy which is generalized to the relativistic energies E expressed in terms of the kinetic energies and the potential energies of particles which has the ability of interconversion between the mass and energy such that:

$$E^2 = p^2c^2 + m^2c^4 \equiv \hat{H}^2.$$

This equation has two values of energies as solutions. Solution of this equation gives two values of energies; positive and negative. Positive energies, obtained as a solution of this equation, correspond to particle energies as classical solutions of this equation. One positive energy solution is the ordinary total energy solution for particles, but negative energy solutions in relativistic quantum mechanics are identified as antiparticles. This simply doubles the particle sector in quantum field theory (QFT). These negative energy solutions may determine the outgoing states which reduce the mass or potential energy of a state. This description of the antiparticle is used to understand the concept at low energy or in non-relativistic quantum mechanics. However, the existence of antiparticles as real particles is now well-tested at relativistic energies. This simply expands the particle section to double in four-dimensional formalism. It simply shows that the expansion of coordinates is always associated with the expansion of the particle sector.

doi:10.1088/978-0-7503-6054-8ch8

8.2 Relativistic generalization of the Schrödinger equation

The notion of particles and antiparticles provides a method to write a complete wavefunction with four-components combining particles and antiparticles in a single state. The fermionic states are composed of two-dimensional particles and antiparticles as two spinors. In this way, the four-dimensional formalism combines relativity and quantum mechanics together. In addition to incorporating antiparticles, we need to introduce spin to develop the four-dimensional generalizations. Therefore, the spin statistics play a crucial role in the relativistic generalization of the Schrödinger equation. However, we have to develop separate equations for bosons and fermions, which are identified as the Klein–Gordon equation or the Dirac equation corresponding to the particles with integral spins (bosons) and with half-integral spins (fermions), respectively. Understanding the derivation of these equations is what describes the need of spin in relativistic formalism combining the incoming and outgoing states of particles. The relativistic equations of motion at relativistic energies lead to the quantization of fields and mostly are applied in the rest frame of the individual particle at quantum scale.

8.2.1 Klein–Gordon equation

The Klein–Gordon equation (K–G equation) is used as an equation of motion of bosons and the spin statistics of bosons does not impose any limit on the number of particles in a single state. Starting with the relativistic equation of energy and substituting the energy and momentum operators from non-relativistic quantum mechanics allows us to write this equation in the form of four-vectors. This gives the K–G equation in four-dimensional space–time or energy–momentum coordinates. For this purpose, we use $\hat{p}_\mu = -\iota\hat{\partial}_\mu$ for $\hbar = c = k_B = 1$ in Euclidean space by using the three-momentum operator as:

$$\vec{\hat{p}} = -i\hbar\vec{\nabla} \equiv -i\hbar\hat{\partial}_i$$

and the energy operator as:

$$\hat{E} = i\frac{\partial}{\partial_t} \equiv i\hat{\partial}_t$$

for $c = 1$. These operators are then rearranged in a relativistic equation as:

$$p^2 - \frac{E^2}{c^2} = (mc)^2$$

which gives a simple form of the K–G equation. In this equation, particles increase energy when added to a system and decrease energy when they are removed, with relativistic energies and mass related through the Einstein's equation for potential energy $E = mc^2$ giving a common solution for positive and negative energy states.

The K–G equation describes the dynamics of relativistically-moving scalar particles of mass m as:

$$-\partial_\mu\partial^\mu\phi = m^2\phi$$

such that ϕ is a wavefunction of scalar particles and the square of total kinetic energy operator $\partial_\mu \partial^\mu$ gives the square of the particle's mass which corresponds to potential energy of a relativistic mass for $c = 1$. The constant mass can then be interpreted as the conservation of four-momentum and this mass changes its sign for antiparticles. Now if ϕ represents an incoming particle (or outgoing antiparticle) with mass m, then $\phi^\dagger$ corresponds to an incoming antiparticle (or outgoing particle) with the same mass. The K–G equation is a combined equation in the four-momentum (energy–momentum) formalism which gives two solutions as a scalar doublet of say ϕ_1 and ϕ_2. An equation uses ϕ_1 as a particle solution with a positive mass and ϕ_2 as antiparticles with a negative mass. Einstein's equation can actually be rewritten using this new four-dimensional d'Alembertian operator written as:

$$\Box \equiv \partial_\mu \partial^\mu = (\partial_i \partial^i - \partial_t^2) = \frac{\partial^2}{\partial x_1^2} + \frac{\partial^2}{\partial x_2^2} + \frac{\partial^2}{\partial x_3^2} + \frac{\partial^2}{\partial x_4^2}$$

for $\hbar = c = 1$, where (x_1, x_2, x_3, x_4) are the four coordinates of the space–time coordinate system. A four-vector formalism is written in the form of d'Alembertian operator as a single K–G equation:

$$(\Box + m^2)\phi = 0, \tag{8.1}$$

namely the K–G equation, which simply reproduces Einstein's equation $E = mc^2$ for a particle at rest. We will not define a massless and spinless scalar in nature. However, if we had one, the K–G equation for massless scalars would attain the form $\partial_\mu \partial^\mu \phi \equiv \Box\, \phi = 0$, which will not contribute any energy to a system at rest. Hence it cannot be a real particle.

The K–G equation typically describes the dynamics of spinless scalars and the solution of this equation is represented as a four-component scalar field which represents scalar particles such as Higgs and other massive scalars which may have to be added to a theory in the effort of combining various interactions in the framework of group theory. The existence of Higgs is associated with the unification of gauge theories. In these gauge theories, the interaction is described as Yukawa theory and its mediators are gauge bosons corresponding to every gauge theory. Photons, $W^\pm$, z^0 and gluons are all examples of gauge bosons. Most of the gauge bosons are massless except $W^\pm$, z^0. The masslessness of a photon is a requirement of gauge invariance which makes it an antiparticle of itself just as indicated in the K–G equation for massless scalars. However, the photon field is a vector field so we develop special equations for photons due to their special characteristics.

Mediators are vector bosons and every vector particle has an associated vector field. The photon with its own vector field is described as a vector potential in three-dimensional space which describes photons as the quanta of energy of electromagnetic signals. The electromagnetic waves combined with electric and magnetic fields help to develop Maxwell's equations. This formalism allows us to rewrite a compact form of all four Maxwell equations in Cartesian coordinates as a single equation:

$$\partial^\mu F_{\mu\nu} = \Box\, A^\nu - \partial^\nu(\partial_\mu A^\mu), \tag{8.2}$$

The K–G equation can fully describe the dynamics of spinless scalar bosons and the solution of this equation is represented as a four-component scalar field. Vector nature is induced by the charge of a particle. This equation includes scalar particles such as Higgs and other massive scalars which appear in various interaction theories and their existence depends on the models of unification theories. During the unification of gauge theories, high energy physicists are compelled to extend the particle sector which is only possible by adding unknown massive scalar particles which disappear at low energies due to breaking of the gauge symmetries. These new scalars are added by hand for extended dimensions of unified theories. Unitary groups provide frameworks for the extension of coordinate space along with the scalar particles themselves. At this point, high energy physics becomes too technical to be discussed here. Additionally, special mathematical tools are required to understand it. This discussion is out of the scope of this book and hence postponed for now.

8.2.2 Equation of motion of vectors

Spin-zero particles, like photons, may have a specific direction and are indicated as spin-1 vector bosons. They satisfy the K–G equation but its solution is not a scalar field. The vector particle solution as a four-vector A_μ corresponds to a four-dimensional field with three components of electric field ($\overrightarrow{E}$) and one imaginary magnetic field ($\overrightarrow{B}$) component such that $A_\mu = (\overrightarrow{E};i\overrightarrow{B}/c)$. The vector particle may or may not have a mass m. QED and quantum chromodynamics (QCD) have massless vector bosons as photons or gluons, whereas the electroweak theory is mediated by massive gauge bosons as (W^+, W^- and Z^0). We mainly consider a photon field which is only a vector particle in QED. The antisymmetric nature of this vector field allows us to define a transformation matrix as:

$$F_{\mu\nu} = \partial_\mu A^\nu - \partial^\nu A_\mu$$

and all the Maxwell's equations can be expressed by a single equation:

$$\partial^\mu F_{\mu\nu} = 0,$$

which obviously summarizes the most important laws of classical electrodynamics including Ampère's law, Faraday's Law, and Gauss's Law using electric and magnetic fields. These laws play a fundamental role in describing the dynamics of electromagnetically-interacting systems at macroscopic level and inevitably contribute to the implementation of dynamics of charge in modern technology, integrating electric and magnetic fields along with electromagnetic radiation. The antisymmetric function ε_μ ensures the antisymmetric nature of the photon field, describing the vector nature of a photon.

However, it is worth mentioning for completion that the given set of equations is applicable with the relevant modifications due to the nature of various interactions and they are used to describe a gauge theory. Description of an interaction theory in the form of a gauge theory is related to the vector nature of the mediators of a gauge theory. The nature of interaction and the corresponding symmetries of a gauge theory are represented by the corresponding spin-1 fields (states) of the mediators (vector bosons) like photons (electrodynamics), the electroweak interaction mediators (γ, W^+, W^- and Z^0) and gluons as mediators of QCD. There are other possible vector bosons which are created as short-lived intermediate states at very high energies and are usually model-dependent. Resonances are examples of such states. However, a detailed discussion of high-energy gauge theories such as QCD and electroweak theories is out of the scope of this book. Moreover, it is worth mentioning that a true gauge theory of gravity is still not fully developed. Since we cannot yet formulate a successful gauge theory of gravity at the quantum scale which is fully applicable to both microscopic and macroscopic level, we do not fully understand the nature of the mediator of gravity, if it is possible to exist. As of now, we model the mediator of gravity as a mediator with spin 2.

8.2.3 Equation of motion of photons

We now generalize the K–G equation for vector particles like photons (vector bosons) to obtain vector fields as a solution of the K–G equation and describe the corresponding interaction theory in four-dimensional formalism. We need to incorporate the direction of motion of the corresponding vector field which describes the direction of the corresponding interaction as well. Then we write a four-dimensional vector field as:

$$A_\mu = \varepsilon_\mu e^{ik\cdot r}$$

where ε_μ indicates the four-dimensional antisymmetric vector representing a block representation of two-dimensional particle and two-dimensional antiparticle states comprised of incoming and outgoing states. This four-vector field A_μ corresponds to a four-dimensional field with three electric field components and one imaginary magnetic field component such that $A_\mu = (\overrightarrow{E};i\overrightarrow{B}/c)$. The direction of B is dependent on the direction of E and is not treated as an independent variable. The direction of E determines the direction of the polarization vector. Remember that the electric and magnetic fields are both vectors which are related in Cartesian coordinates as:

$$\overrightarrow{B}\propto(\overrightarrow{\nabla}\times\overrightarrow{A}).$$

Its analogue in classical electrodynamics can be seen in terms of the three-dimensional Cartesian vector field $\overrightarrow{A}$ and a corresponding scalar electromagnetic field ϕ. This expression of Maxwell's equations provides a missing link between classical electrodynamics and the relativistic version of electrodynamics ensuring the application of fundamental laws of electrodynamics to relativistic quantum mechanics of electrodynamics known as QED.

8.2.4 Dirac equation

The most generalized relativistic form of the equations of motion of fermions is developed combining the regular Schrödinger equation of quantum mechanics and the total energy equation of relativity along with the fermions spin statistics. It is not a straightforward generalization such as the K–G equation. The spin statistics of fermions at relativistic energies is called Fermi–Dirac statistics. It gives a block representation of two-component wavefunctions of spin 1/2 and spin –1/2 basis vectors which can provide a set of four-components basis vectors. Two-dimensional wavefunctions with incoming and outgoing states are replaced by a four-dimensional wavefunction which includes two spin vectors altogether, which can be expressed in two doublets corresponding to spin half fermions (particles and antiparticles). This formalism was formulated by Paul Dirac in 1928, and it plays a fundamental role in the development of QFT. Its derivation is also based on generalization of the Schrödinger equation in the matrix form. Now, we start writing the relativistic equation in natural units ($c = \hbar = 1$) as:

$$E^2 = |\overline{p}\ |^2 + m^2 \tag{8.3}$$

Using the operator formalism, it leads to the equation:

$$\frac{\partial^2 \psi}{\partial t^2} = \overrightarrow{\nabla}^2\ \psi - m^2\psi \tag{8.4}$$

This is a second-order equation in spatial as well as time derivatives. For the plane-wave solutions:

$$-E^2\psi = -|\overrightarrow{p}\ |^2\psi - m^2\psi$$

such that:

$$E = \pm\sqrt{|\overrightarrow{p}\ |^2 + m^2} \tag{8.5}$$

However, the fermions can contribute to either positive or negative energy states and may have two spin states as well. The same argument of particles and antiparticles is not so simple for fermions. Particle and antiparticle substitute each other, but the addition and subtraction in finding the probability of particles cannot be accepted directly. This problem was realized by Dirac and motivated him to incorporate fermion spin into the particle (incoming) and antiparticle (outgoing particle) wavefunctions. Two-component individual spinors were introduced as $u(p)$ for particles and $v(p)$ for antiparticles as solutions of individual equations of motion for fermions in quantum field theory. Two separate spinors were needed to solve two different equations, one for particles and the other for antiparticles. In addition, two-dimensional spin operators were associated with spinors to distinguish their spin contribution to the total spin of the system. A combination of two-dimensional $u(p)$ and $v(p)$ together provided a four-dimensional wavefunction represented as a block representation of particle-antiparticle doublet.

Dirac's proposed solution and not only utilized $u(p)$ and $v(p)$ (corresponding to particle and antiparticle states), but also two-dimensional Pauli spin matrices. Pauli spin matrices are only two-dimensional. Dirac used them to develop four-dimensional wavefunctions and spin operators to incorporate particles and anti-particles together. He proposed a common solution as:

$$\hat{H}\psi = (\overrightarrow{\alpha}\cdot\overrightarrow{p} + \beta m)\psi = i\frac{\partial\psi}{\partial t} \tag{8.6}$$

where $\hat{H}$ is the Hamiltonian operator. Expanding it:

$$\left(-i\alpha_x\frac{\partial}{\partial x} - i\alpha_y\frac{\partial}{\partial y} - i\alpha_z\frac{\partial}{\partial z} + \beta m\right)\psi = i\frac{\partial}{\partial t}\psi$$

A free particle with a mass m must always satisfy the condition $E^2 = \overline{p}^2 + m^2$. However, the Dirac equation has to be consistent with the K–G equation but still satisfy the properties of these equations as:

$$\alpha_x^2 = \alpha_y^2 = \alpha_z^2 = \beta^2 = 1$$

$$\alpha_j\beta + \beta\alpha_j = 0$$

$$\alpha_j\alpha_k + \alpha_k\alpha_j = 0, \ (j \neq k)$$

such that the block representation of these matrices is that:

$$\beta = \begin{pmatrix} I & 0 \\ 0 & -I \end{pmatrix}, \qquad \alpha_j = \begin{pmatrix} 0 & \sigma_j \\ \sigma_j & 0 \end{pmatrix}$$

where three components of σ_i from quantum mechanics along with a 2×2 identity matrix gives a set of four-vectors given as:

$$I = \begin{pmatrix} 1 & 0 \\ 0 & 1 \end{pmatrix}, \ \sigma_x = \begin{pmatrix} 0 & 1 \\ 1 & 0 \end{pmatrix}, \ \sigma_y = \begin{pmatrix} 0 & -i \\ i & 0 \end{pmatrix}, \ \sigma_z = \begin{pmatrix} 1 & 0 \\ 0 & -1 \end{pmatrix}$$

These conditions require α_j and β to be four mutually anti-commuting matrices. The wavefunction must be four-component Dirac spinors:

$$\psi = \begin{pmatrix} \psi_1 \\ \psi_2 \\ \psi_3 \\ \psi_4 \end{pmatrix}.$$

A consequence of the first-order equations in spatial and time coordinates is that the wavefunction has new degrees of freedom. This is how we import three-dimensional Pauli spin matrices into Dirac matrices (α and β) to introduce the spin statistics which works for both particles and antiparticles altogether. The Pauli spin matrices were considered to write the four anti-commuting matrices. The probability density is then expressed by introducing a compact representation of matrices to write a single equation in four-dimensional space incorporating the Fermi statistics expressed as the Dirac equation which satisfies the basic conditions of probability density:

$$\rho = \psi^\dagger\psi$$

$$\rho = |\psi_1|^2 + |\psi_2|^2 + |\psi_3|^2 + |\psi_4|^2$$

The Dirac equation has probability densities that are always positive. The solutions to the Dirac equation are Dirac spinors which give rise to the intrinsic spin property of the particles. To rewrite it in a relatively general way with particular properties, let us introduce Dirac gamma matrices:

$$\gamma^0 \equiv \beta;\ \gamma^1 \equiv \beta\alpha_x;\ \gamma^2 \equiv \beta\alpha_y;\ \gamma^3 \equiv \beta\alpha_z$$

such that the Dirac equation can be written as a single equation in compact form as:

$$i\gamma^1\frac{\partial\psi}{\partial x} + i\gamma^2\frac{\partial\psi}{\partial y} + i\gamma^3\frac{\partial\psi}{\partial z} + m\psi = -i\gamma^0\frac{\partial\psi}{\partial t}$$

This can be expressed as a compact equation as:

$$(i\gamma^\mu\partial_\mu - m)\psi = 0, \tag{8.7}$$

which is a four-dimensional version of the Schrödinger equation which can fully describe relativistic quantum mechanics. In this equation, γ^μs are commonly known as gamma matrices of QFT. ∂_μ denotes the four-dimensional gradient, m is the rest mass of a fermion, and ψ is a wavefunction, which in this case is a four-component wavefunction which is made up of a block of two-dimensional spinors. Now the wavefunction ψ is a four-dimensional column vector ψ in a block representation of particle u and antiparticle v doublets. These spinors describe the spin degrees of freedom for fermions, which are essential for the correct relativistic treatment of spin-1/2 particles such that the complete wavefunction can be given as:

$$\psi = u(x)e^{-ipx/\hbar} + v(x)e^{ipx/\hbar}, \tag{8.8}$$

Another notation is commonly used here that we define as:

$$i\gamma^\mu\partial_\mu = \gamma^\mu p_\mu \equiv i\not{p}$$

and a combined Dirac equation in four-dimensional space is written as:

$$(\not{p} - m)\psi = 0$$

which can even be decomposed a two-dimensional equation for particles and antiparticles separately:

$$\psi = \begin{pmatrix} u(p) \\ v(p) \end{pmatrix} = \begin{pmatrix} particle \\ antiparticle \end{pmatrix}$$

such that the two different equations that are satisfied by two incoming and outgoing particles are:

$$(\not{p} - m)u(p) = 0 \tag{8.9}$$

and:

$$u^\dagger(p)(\not{p} - m) = 0$$

The corresponding equations for incoming and outgoing antiparticles can be written as:

$$(\not{p} + m)v(p) = 0 \tag{8.10}$$

and:

$$v^{\dagger}(p)(\not{p} + m) = 0$$

Each of these equations has two solutions for each doublet. The set of four different Dirac equations for particles $u(p)$ and antiparticles $v(p)$ correspond to particle and antiparticle spinors [$u(p)$ and $v(p)$], respectively. These spinors further satisfy the conditions:

$$u_r^{\dagger}(p)u_r(p) = v_r^{\dagger}(p)v_r(p) = \frac{E_p}{m}.$$

One of the most significant features of the Dirac equation is that it predicts the existence of antimatter both for fermions and bosons. This equation naturally yields solutions corresponding to both positive and negative energy states. This duality was initially perplexing, but it was later understood to imply the existence of particles identical in mass but opposite in charge to the known fermions. These antiparticles, such as positrons, were later experimentally confirmed, marking a major triumph for the theory. However, later on this equation became a general equation for fermions in any interaction theory and accommodated all types of particles with all types of quantum numbers including additive and multiplicative eigenvalues or even the projection operators.

Gamma matrices have very specific properties and gamma algebra is a special type of matrix algebra which can later be used in every gauge interaction theory in QFT. In particular, it allows us to calculate the decay rates and scattering cross-sections in all fundamental interactions at high energies. Some of the useful well-known properties of gamma matrices are included in the appendix so that a reader can easily use it without going into the proof of identities every time. It therefore provides an effective approach with sufficient mathematical tools which can be applied to any local interaction theories including local short-ranged theories such as weak and strong interactions. In various interaction theories, the participating particles are not all similar in nature and their behavior depends on their intrinsic properties or the associated quantum numbers. They may not always exist as free independent particles due to their participation in short-ranged nuclear interactions only. Quarks provide a good example of such strongly- and weakly-interacting particles with fractional charges which are always found in bound states and these quarks with fractional charges cannot participate in electromagnetic interaction unless they are bound together to give integral charges like protons.

On the other hand, for multiparticle systems, the second quantization (discussed previously) allows us to identify states by the number of particles instead of defining an individual particle state. Such states represent the quantized states that exist in the form of quantized fields instead of the point particles. This is how the idea of quantization of fields is introduced in high-energy physics and developed as QFT. This field quantization allows us to define all quantum states and operators as quantized fields and gives birth to the most generalized interaction (gauge) theories in QFT. This theory shows discreteness in fields associated with particles instead of state variables and the field operators leading to the invariance of the system due to the exchange of identical particles.

Spinless particles have scalar fields and remain invariant in space. Scalar particles may have just one degree of freedom. However, its representation in four-dimensional space is possible even if it has only one nonzero component. It just ensures the invariance of its four-dimensional magnitude, which can be identified as a quantization of the four-dimensional field associated with a particle, and we find a way to represent particles as quantized fields which represent the regions of its interactions. The Dirac equation lets us rewrite the Hamiltonian and Lagrangian of an interaction theory and the kinetic energy and the potential energy or both of them are now expressed in four-dimensional space.

8.3 Development of quantum field theory

Difference in the spin statistics of fermions and bosons led to the derivation of two different equations of motion for fermions and bosons. Their dynamic behavior is not the same either. The kinetic energy of fermions and bosons is then inserted in the Lagrangian as two separate terms, and the different terms are used to indicate interactions between fermions and bosons. Once a complete Lagrangian of a theory is written, it has to be proven invariant even if certain transformations of coordinates or the corresponding fields are changed. These transformations which keep the Lagrangian and then the corresponding action unchanged are called gauge invariant. The basic principle which requires the invariance of the Lagrangian for a physically acceptable interaction theory is called gauge theory. An interaction theory cannot be accepted as a correct physical theory until it satisfies the gauge invariance.

8.3.1 Gauge theories

Gauge invariance basically requires that the fundamental nature of the corresponding interaction is preserved by the assurance of the invariance of Lagrangian if certain transformations take place. These transformations are called gauge transformations. The invariance of the Lagrangian under gauge transformations is called gauge invariance and the conservation rules obtained as a result of these transformations are called gauge transformations. This invariance of the Lagrangian is used to determine the various conservation principles of interaction theories and allowed transformations of the Lagrangian, which does not affect the behavior of the interaction or dynamics of a system.

Lagrangian formalism helps to determine the conservation rules of the corresponding theory. All the particles with certain required characteristics can realize or participate in the same interaction if they can follow the required conservation rules obtained from the gauge invariance. Then the unitarity of the transformation matrices requires the Lagrangian to be invariant under certain transformations. These transformations make a group and the generator of this group determines the number of mediators of the interaction, and the matrix elements of the group determine the number of particles needed to make a complete set of the group which can participate in the given interaction. All these conditions are needed for a physical theory. Every interaction is only a physical interaction if it is gauge invariant and all the fundamental interactions should be represented in terms of the gauge group. This gauge group provides a set of transformations which satisfy the gauge invariance. The generators of these groups correspond to mediators of a theory which are also identified as gauge particles like photons and gluons. Mediators are responsible for the interaction and connect various elements of the gauge group or indicate the transition among various elements managed by the interactions between particles which indicates the probability of production of various particles.

8.3.2 Lagrangian formalism in QFT

QFT is a generalized form of gauge theories which is locally defined at a very small scale which could be applicable to nuclear interactions at higher energies. Quantum field theories use local gauge theories to develop the four-dimensional Lagrangian formalism. This local form of the Lagrangian is defined at a given point of space and time which corresponds to the action at a given time at a certain point in space for a rapidly changing energy. This is called Lagrangian density ($\mathscr{L}$) in four-dimensional space. The classical Hamiltonian

$$H = p_i(t)\dot{q}_i(t) - L$$

of such a system locally attains the form of $\mathscr{H}$ at a given instant of time t which is written as:

$$\mathscr{H} = \Sigma_i p_i(t)\dot{q}_i(t) - \mathscr{L}.$$

In QFT theories, we do not need to quantize the spatial variables as measurement in space becomes difficult due to the wave nature of particles and the momentum space is considered to be more reasonable for measurements. QFT works more in the Fourier-transformed coordinates which could be converted into real space–time coordinates at any point through the inverse Fourier transform. The idea of generalized coordinates allows us to discuss the field quantization, such that the total Lagrangian is defined as:

$$L = \int \mathscr{L} d^3 q_i$$

for the action due to the scalar field in terms of generalized coordinates q_i instead of the spatial or momentum coordinates. We define the quantized field in space as $\phi(q_i, t)$. Let us define the canonical conjugate (or conjugate) momentum in space as:

$$\pi(q_i, t) := \frac{\partial \mathscr{L}}{\partial \dot{\phi}(q_i, t)} \tag{8.11}$$

The field $\phi(q_i, t) \equiv \phi(x)$ now depends on x (a four-dimensional space–time coordinate) and is considered as a local field which is required by the field quantization. The corresponding Hamiltonian density $\mathscr{H}$ is then defined in terms of the Lagrangian density $\mathscr{L}$ which gives the total Hamiltonian as:

$$H = \int d^3q_i \mathscr{H}$$
$$= \int d^3q_i [\pi(t, q_i)\dot{\phi}(t, q_i) - \mathscr{L}]$$
$$H = \Sigma_i q_i \dot{q}_i - L$$

such that:

$$\pi(t, q_i) = \dot{\phi}(t, q_i) \equiv \pi(q_\mu)$$

where q_μ are the four-dimensional generalized coordinates (q_i, t) such that q becomes a four-dimensional variable just like the four-dimensional x. Therefore, the total Hamiltonian of a system can be written as:

$$H = \int d^3q_i \left[\dot{\phi}^2(q_\mu) - \frac{1}{2}(\partial_\mu\phi)(\partial^\mu\phi) + \frac{1}{2}m^2\phi^2(q_\mu)\right]$$

$$H = \int d^4q_\mu \left[\frac{1}{2}(\Box\phi(q_\mu))^2 + \frac{1}{2}m^2\phi^2(q_\mu))\right] \tag{8.12}$$

The time independent form of this equation describes the corresponding dependence on three-dimensional generalized coordinates ϕ_μ and π_μ, which correspond to stationary states and are self-adjoint with canonical commutation relations:

$$[\phi(q_{1i}), \pi(q_{2i})] = i\delta^{(3)}(q_{1i} - q_{2i}), \ [\phi(q_{1i}), \phi(q_{2i})] = 0 = [\pi(q_{1i}), \pi(q_{2i})] \tag{8.13}$$

Using Fourier transformation on the fields as:

$$\phi(q_i) = \int \frac{d^3p_i}{(2\pi)^3}\tilde{\phi}(p_i)e^{ip_i q^i} \tag{8.14}$$

$$\pi(q_i) = \int \frac{d^3p_i}{(2\pi)^3}\tilde{\pi}(p_i)e^{ip_i q^i} \tag{8.15}$$

where $\phi^\dagger(q_i) = \phi(-q_i)$. To compute H in Fourier space, we use these expressions as:

$$\frac{1}{2}\int d^3q_i(\nabla\phi(q_i))^2 = \frac{1}{2}\int d^3q_i\left(\int \frac{d^3p_i}{(2\pi)^3} \nabla \ e^{ip_i q_i}\tilde{\phi}(q_i)\right)^2$$

$$\frac{1}{2}\int d^3q_i(\nabla\phi(q_i))^2 = \frac{1}{2}\int d^3q_i\left(\int \frac{d^3p_i d^3q_i}{(2\pi)^3} \nabla \ e^{ip_i q_i}\phi(q_i)\right)^2$$

The four-dimensional version of kinetic and potential energies gives the complete Lagrangian in relativistic quantum mechanics in terms of four-dimensional generalized coordinates q_μ and their conjugate momenta $p_\mu = \pm i\hbar\frac{\partial}{\partial q^\mu} \equiv \pm iD_\mu$. An interaction theory is then written in the form of generalized coordinates as:

$$L_{\text{QFT}} = i\hbar c\overline{\psi}\gamma^\mu D_\mu\psi - mc^2\overline{\psi}\psi - V(x_\mu) \tag{8.16}$$

where the simplest form of the four-dimensional generalization represented by $q_\mu = (\vec{x}, ict)$ or its conjugate moment $p_\mu = (\vec{p}, iE/c)$ starting with a scalar stationary field $\phi(x_\mu)$ denoted and obeys the K–G equation which is derived from the Schrödinger equation. Four-dimensional Lagrangian formalism is identified as canonical formalism as it incorporates the canonical transformation and can lead to finding the conservation rules of the theory using equation (8.11). For that purpose, let us switch from the Lagrange formalism to canonical formalism of the classical theory, which can include all transformations conserving the Hamiltonian. These canonical transformations keep the action conserved as well. The canonical momentum conjugate to generalized coordinates (q_i) is written as:

$$p_i(t) = \frac{\partial L}{\partial q_i}$$

which can be generalized to:

$$p_\mu = \frac{\partial \mathscr{L}}{\partial q_\mu}$$

where $\mathscr{L}$ corresponds to the Lagrangian density because in four-dimensional formalism, the particle fields $\phi(x)$, $\psi(x)$ appear to be local fields, whereas the classical Hamilton $H = \Sigma_i p_i(t)\dot{q}_i(t) - L$ into the Legendre transformation of the Lagrange function is:

$$\mathscr{H} = \Sigma_i p_i(t)\dot{q}_i(t) - \mathscr{L}$$

Kinetic energy is calculated in the same way from each interaction theory and is related to the mass of interacting particles and their dynamics. However, the potential representing each fundamental interaction depends on the intrinsic properties of particles and the range and nature of the interaction and their strength and its behavior (as well as symmetries). We will discuss QED formalism in detail to describe the local form of QFT in electrodynamics at the quantum scale at relativistic energies.

Classical Lagrangian theory mainly deals with gravitational interactions and works on the macroscopic scale at large distances which covers the scope of gravity. QFT, on the other hand, addresses the problems at microscopic level or can be generalized to the individual particle scale. This general formalism therefore addresses all the fundamental interactions described using quantum theory, or more precisely in the most general form of QFT. This QFT is developed in a general way to accommodate all interaction theories such that even the local Lagrangian has to be written as:

$$\mathscr{L} = \mathscr{L}_0 + \mathscr{L}_I$$

and the Hamiltonian as:

$$\mathscr{H} = \mathscr{H}_{\mathscr{I}} + \mathscr{H}_{\mathscr{J}}$$

at relativistic energies. $\mathscr{L}$ and $\mathscr{H}$ can accommodate all the information about the corresponding interaction in terms of its potential. A canonical transformation theory gives the information about the fundamental interactions and a general formalism is obtained to be able to accommodate all fundamental interactions.

8.3.3 Quantum field theory

Hideki Yukawa initially developed a theory to describe nuclear interaction as an interaction between fermions, taking place through the exchange of undetectable scalars. The exchange of a virtual boson or a gauge boson is required to describe an interaction between fermions, which has been generalized as a gauge theory and gauge bosons are represented as the generator of a group mathematically. We will concentrate on QED in this book mainly as the most successful gauge theory due to its unique ability to be generally applicable from individual particles to the cosmic scale.

Electromagnetic interactions are realized due to the exchange of virtual photons and are represented by the vector fields A_{μ}. Photon fields are quantized as quanta of energy which have all the characteristics of particles which have been found to be massless to maintain the gauge invariance, which is a basic requirement of any physically acceptable gauge theory. However, photons are treated as massless particles during the study of interaction processes. A virtual exchange of the photon field in a way describes the electromagnetic interaction. These mediators are responsible for every gauge interaction theory and generate a complete set of the unitary operators of an interaction which has to be unitary operators for conservative forces of a gauge theory. These unitary operators make a group which is a complete set of unitary matrices which satisfy the criteria of a group. The exchanging particles in interactions virtually were mathematically incorporated as vector fields (e.g., A_{μ}) in the Lagrangian and are identified as propagators (of the mediators).

Group theoretic representation of interaction theories miraculously provided a complete description of fundamental interactions as gauge theories with their symmetries and conservation rules. Gauge invariance is a mechanism which identifies the symmetries of the Lagrangian and the associated conservation laws of a theory. The gauge transformations along with the invariance of the Lagrangian and the relevant conservation rules altogether help us to understand the hidden features of every fundamental interaction. One of the greatest outcomes of gauge theories is to develop models of unification of gauge theories. Among them the electroweak theory as a unified gauge theory has been now tested and work is ongoing to test the standard model. However, a direct experimental verification seems to be out of the scope of existing technologies.

Electromagnetic current is associated with the flow of charged particles. The gauge current is a generalization of the same concept, which is not necessarily related to charged particles only. Gauge current density in defined as the difference between the net flow of particles between the incoming and outgoing states. The total current J^μ is defined in terms of the probability of flow of a particle j^μ and is related as $J^\mu \equiv Nj^\mu$ by a comparison between the probability of inflow and outflow of charged particles and is determined as:

$$j^\mu \equiv Nj^\mu = \iota eN(\phi^* \partial^\mu \phi - \phi \partial^\mu \phi^*) \tag{8.17}$$

The total current can then be obtained by the integration of the current density and the rate of flow of each particle is multiplied by the total number of scalar particles. Scalar particles have no restrictions on the number of particles occupied by the same state. Net current is the probability of flow of current in a particular direction multiplied by the total number of particles in the same state. However, this definition cannot be applied to the charged fermions where each particle has to occupy a different state. The current produced by fermion fields is then written as:

$$J^\mu = \iota e(\overline{\psi}\gamma^\mu\psi - \psi\gamma^\mu\overline{\psi}) \tag{8.18}$$

The rate of flow of a stream of electromagnetically-interacting charged particles (or current) and the interaction of current with charges is also discussed in electrodynamics within the classical range of energies. Classical electrodynamics deals with electric and magnetic fields as the continuous classical fields that are associated with moving charges and satisfy the wave equation. The combined form of antisymmetric classical electromagnetic fields is given as $F_{\mu\nu}$. This classical field matrix is directly borrowed in QED and is perfectly relating the classical and QED making QED the most successful and observable fundamental interaction. Classical electrodynamics works perfectly at the macroscopic level with the flow of current through the medium. Study of the mediating field corresponds to a fundamental interaction. Exchange of a field between two interacting particles is then represented by a propagator which can be considered as a mathematical description of fundamental interactions. We can use a general term of Feynman propagator to understand the nature of interaction in terms of the associated fields. Obviously, we can find all the conservation rules for an interaction theory using the gauge invariance of the Lagrangian of the theory.

8.3.4 Feynman propagator

Interaction between two particles is mathematically interpreted as a response of a particle due to the potential of the interacting particle that is named as the propagator representing the transfer of mediator from one particle to another particle. In normal space–time coordinates, a Feynman propagator $D_F(x - y)$, describes the propagation of a scalar particle from point y to point x as a two-point function of the quantum fields which describes an incoming state and an outgoing state relative to the observation point and is mathematically represented as a retarded Green's function mathematically. By definition, the retarded Green's function is distinguished from the advanced Green's function by a comparison due to different time ordering. The advanced Green's function basically gives the information of the interaction which the incoming state had come across before detection, whereas the retarded one will tell us about what interaction potential is expected to be experienced. The propagator theory is associated by virtual exchange of a scalar or vector boson or a fermion and is represented by Green's function mathematically. A scalar boson propagator is then given as:

$$D_F(x - y) = \langle 0|T(\hat{\phi}(x)\hat{\phi}(y))|0\rangle, \tag{8.19}$$

where T is the time-ordering operator that orders the fields according to their time arguments to keep it a retarded Green's function and gives assurance that the Green's function truly indicates the response to (or effect of) a potential. The advanced Green's function is the one which is not relevant. The T operator therefore is defined to automatically arrange the fields to see the response and ignore the irrelevant part such that:

$$T(\hat{\phi}(x)\hat{\phi}(y)) = \begin{cases} \hat{\phi}(x)\hat{\phi}(y) & \text{if} \quad t_x > t_y, \\ \hat{\phi}(y)\hat{\phi}(x) & \text{if} \quad t_y > t_x. \end{cases} \tag{8.20}$$

In the standard notation of QFT, $|0\rangle$ represents the vacuum state (with no particle in it) and $\hat{\phi}(x)$ is the quantized scalar field operator. The propagator $D_F(x - y)$ gives the probability amplitude for a scalar particle to propagate from point $y^\mu = (t_y, \mathbf{y})$ to point $x^\mu = (t_x, \mathbf{x})$. This concept of virtual exchange of a particle can be expressed in terms of raising and lowering operators as a scalar field (particle) leaves its original state and moves to another state. It indicates that the scalar field changes when an incoming state is added or an outgoing state is stopped by increasing the number of particles in the initial states. An increase in the number of particles by one in the initial states with the field operator $\hat{\phi}(x)$ in terms of the corresponding creation and annihilation operators can be written as:

$$\hat{\phi}(\overline{x}, t) = \int \frac{d^3p}{(2\pi)^{3/2}} \frac{1}{\sqrt{2\omega_{\overline{p}}}} (a(\overline{p})e^{i\overline{p}\cdot\overline{x} - i\omega_{\overline{p}}t} + a^\dagger(\overline{p})e^{-i\overline{p}\cdot\overline{x} + i\omega_{\overline{p}}t}), \tag{8.21}$$

The time-ordered product of two field operators in a vacuum is known as the vacuum expectation value (VEV) of the field and is defined as:

$$\langle 0|T(\hat{\phi}(x)\hat{\phi}(y))|0\rangle = \int \frac{d^4p}{(2\pi)^4} \frac{i}{p^2 - m^2 + i\varepsilon} e^{-ip\cdot(x-y)}. \tag{8.22}$$

This is the Feynman propagator in momentum space. $D_F(x - y)$ is the space–time representation of a scalar particle propagator, whereas its VEVs can be considered in a momentum representation of the same propagator as VEV is mathematically a field theory of $D_F(x - y)$ in momentum space.

Considering $t_y > t_x$, we have:

$$\langle 0|\hat{\phi}(y)\hat{\phi}(x)|0\rangle = \int \frac{d^3p}{(2\pi)^3} \frac{1}{2\omega_{\overline{p}}} e^{i\overline{p}\cdot(\overline{x} - \overline{y}) - i\omega_{\overline{p}}(t_y - t_x)}. \tag{8.23}$$

whereas, for $t_x > t_y$, it gives:

$$\langle 0|\hat{\phi}(x)\hat{\phi}(y)|0\rangle = \int \frac{d^3p}{(2\pi)^3} \frac{1}{2\omega_{\overline{p}}} e^{i\overline{p}\cdot(\overline{x}-\overline{y})-i\omega_{\overline{p}}(t_x - t_y)}. \tag{8.24}$$

Then the general form of the time-ordered propagator in coordinate space as a Fourier transform is written as:

$$D_F(x - y) = \int \frac{d^3p}{(2\pi)^3} \frac{1}{2\omega_{\overline{p}}} (e^{i\overline{p}\cdot(\overline{x}-\overline{y})-i\omega_{\overline{p}}|t_x - t_y|}). \tag{8.25}$$

A four-dimensional generalization of the three-dimensional propagator relates the propagator in four-dimensional space–time coordinates with the energy–momentum propagator, expressed as:

$$D_F(x - y) = \int \frac{d^4p}{(2\pi)^4} \frac{\imath}{p^2 - m^2 + i\varepsilon} e^{-ip\cdot(x-y)}. \tag{8.26}$$

The direction of momentum along with the positive or negative values of energy in four-momentum representation is maintained by time ordering of four-momentum. The standard Feynman propagator in position coordinates is then represented by equation (8.26). The $\imath\varepsilon$ term in the denominator ensures the correct boundary conditions for the causal propagation of particles, avoiding divergences by pushing the poles in the propagator slightly off the real axis. This Feynman propagator can be considered as a standard form of a retarded Green's function between incoming and outgoing states. $D_F(x - y)$ represents the amplitude for a particle to travel from space–time point $y^\mu = (t_y, \overline{y})$ to $x^\mu = (t_x, \overline{x})$.

The Feynman propagator $D_F(x - y)$ as a Green's function for the K–G equation for a free scalar field $\phi(x)$ with mass m can also be expressed as:

$$D_F(x - y) = \int \frac{d^4p}{(2\pi)^4} \frac{i}{p^2 - m^2 + i\varepsilon} e^{-ip\cdot(x-y)}. \tag{8.27}$$

which provides an integral representation of Green's function as a solution of the K–G equation for a source potential given in terms of a Dirac delta function such that the following operation is satisfied as:

$$(\Box + m^2)D_F(x - y) = -\delta^{(4)}(x - y). \tag{8.28}$$

However, the QFT works better in momentum–energy coordinates. A natural coordinate system with real spatial coordinates and imaginary time is identified as Euclidean space in imaginary-time formalism, especially in quantum statistical field theories. In contrast, another more efficient coordinate system proposed by Minkowski is called Minkowski space or real-time formalism which can be obtained by simple rotation of space–time coordinates in four-dimensional space (Wick's rotation) and obtains time as a real coordinate and spacial coordinates as imaginary coordinates. Minkowski space is more natural and useful in QFT as for the experimental observation of particle processes, energy becomes an easily measurable parameter as compared to three-momentum and provides more effective tools in the four-dimensional formalism in momentum space. A Fourier transform is used to convert imaginary time into real-time formalism and provides a very effective tool in various applied fields including signal processing and other calculations in electrical engineering. All the four-momentum results can be converted into real-time results through the inverse Fourier transform at any time.

QFT provides a framework to calculate the individual particle processes including decay rates and scattering crossections of all particles. However, for this purpose, we need to extend the formalism to develop a few more rules for all calculations to study the particle processes in QFT. Since we deal with conservative forces, all of the processes are therefore elastic processes and obey the energy–momentum conservation rules. However, at relativistic energies, conservation of three-momentum and total energy cannot be individually applied in QFT. Here we need to incorporate the relativistic energy equation $E^2 = p^2 + m^2$ which is equivalent to a four-momentum expression $p_\mu p^\mu = m$ in natural units. This relativistic relation then allows us to merge the energy–momentum conservation into four-momentum conservation and allows the conversion of energy into mass and vice versa.

The four-momentum conservation rules play a crucial role in developing techniques to estimate the decay rates of particles and scattering cross-sections. Probability of finding fundamental particles in various initial and final quantum states is incorporated with the classical approach for the calculation of decay rates and scattering cross-sections. And the language of probability is employed in the expressions of decay rates and scattering cross-sections as the fundamental particles due to various intrinsic properties can undergo decays and scatterings in various modes obeying various conservation rules. The integration of probability of statistical physics and quantum mechanics with QFT along with relativistic high energy particle physics introduces a notion of probability in particle processes. The probability of a particular process (channel) compared to the probability of other particles is called a branching ratio of the corresponding decay rate or the scattering cross-section. Summation over all the final states provides the lifetime of a particle under consideration. Similarly, the integration over the solid angle after taking the summation over all the incoming states and the average over the final states gives the total scattering cross-section.

Mass–energy conversion is allowed in relativity and lets heavier particles decay into two or more particles with lower mass. However, at high energies, matter and energy can be used interchangeably. This is the only reason that heavier particles can be produced from light ones as long as all the other required conservation rules are obeyed. The scattering and decay processes are not necessarily elastic processes as the initial and final states involved in a process are not always the same. Various factors contribute to the probability of these processes, and we can look at it more carefully. For this purpose, we define a transformation matrix (called S-matrix) which describes the multiparticle processes also and is not required to be a unitary matrix. The S-matrix gives the account of incoming and outgoing states which is a kind of generalization of the wavefunction. The decay mode or scattering channel given in terms of S-matrix describes a process such that the modulus square of the S-matrix gives the probability of that process under given conditions. In general, it needs to be integrated to calculate the probability of the given process.

8.3.5 *S*-matrix

Starting with the regular Hamiltonian H, the full Hamiltonian includes the interaction Hamiltonian as:

$$H = H_0 + H_{\text{int}}. \tag{8.29}$$

The scattering matrix, or S-matrix, connects the incoming and outgoing states of a process by providing a mathematical tool to describe the transition of an initial state into a final state under the effect of the relevant interaction. An S-matrix is an operator that transforms the incoming states $|\psi\rangle$ into an outgoing state $\langle\psi|$ such that the S-matrix gives the probability of transition as:

$$|\langle\psi|S|\psi\rangle|^2.$$

It is formally defined as:

$$S = \lim_{t\to\infty} e^{iH_0 t} e^{-iHt} e^{-iH_0 t}, \tag{8.30}$$

In practice, the elements of the S-matrix give the transition amplitudes between the initial and final states in a scattering process. For example, the probability amplitude for transitioning from an initial state $|\alpha\rangle$ to a final state $|\beta\rangle$ is given by the S-matrix element, defined as:

$$\langle\beta|S|\alpha\rangle = \langle\beta, \text{ out}|\alpha, \text{ in}\rangle. \tag{8.31}$$

The transition between an incoming and an outgoing state could take place in a single step or it may involve several small steps (discrete or continuous). The S-matrix can therefore be expanded in a series called the Dyson series. The perturbation theory demands these steps to be small enough to be represented as a convergent series. If we denote the interaction Hamiltonian in the interaction picture as $H_{\text{int}}(t)$, the Dyson series for S-matrix is written as:

$$S = \mathcal{T} \exp\left(-i\int_{-\infty}^{\infty} dt\ H_{\text{int}}(t)\right), \tag{8.32}$$

where $\mathcal{T}$ represents the time-ordering operator. A perturbative expansion of this expression yields:

$$S = 1 - i\int_{-\infty}^{\infty} dt\ H_{\text{int}}(t) - \frac{1}{2}\int_{-\infty}^{\infty} dt \int_{-\infty}^{\infty} dt'\ \mathcal{T}(H_{\text{int}}(t)H_{\text{int}}(t')) + \cdots. \tag{8.33}$$

The Feynman amplitude $\mathcal{M}$ for the transition from initial state $|i\rangle$ to final state $|f\rangle$ is defined as:

$$S_{fi} = \delta_{fi} + (2\pi)^4\delta^4\left(\sum p_f' - \sum p_i\right)\prod_i \left(\frac{1}{2\pi E_i}\right)^{1/2}\left(\frac{1}{2\pi E_f'}\right)^{1/2} \prod_i (2m_i)^{1/2}\mathcal{M} \tag{8.34}$$

The In and Out states provide a rigorous way to calculate scattering amplitudes in interacting QFTs. By assuming free-particle behavior at $t \to \pm\infty$, we can relate interacting theory calculations to observable quantities. Specifically, the elements of the S-matrix give the probabilities of transition between various particle states, allowing us to calculate cross-sections and decay rates:

$$\sum_f |S_{fi}|^2 = 1 \tag{8.35}$$

$$S = \sum_{n=0}^{\infty}(-1)^n \int_{-\infty}^{\infty} dt_1 \int_{-\infty}^{t_1} dt_2 \cdots \int_{-\infty}^{t(n-1)} dt_n H_1(t_1)H_1(t_2) \cdots H_1(t_n) \tag{8.36}$$

$$S = \sum_{n=0}^{\infty} \frac{(-1)^n}{n!} \int_{-\infty}^{\infty} dt_1 \int_{-\infty}^{\infty} \cdots \int_{-\infty}^{\infty} dt_n T[H_1(t_1)H_1(t_2) \cdots H_1(t_n)] \tag{8.37}$$

8.4 Calculation of decay rates and cross-sections

The time-ordered product gives the assurance that we use propagators as retarded Green's functions and they contribute to the particle processes. However, there is more than one possibility of occurrence of physical processes and various modes. One particle may decay into a group of various modes which are observed through different channels, depending on the project and the available equipment. The identification of decay rates may be due to the initial state or the final product. Similarly, scatterings may be categorized by the mediators corresponding to the initial and final states and the relevant conservation rules.

8.4.1 Decays

All the individual particle decay rates incorporate quantum mechanics at any level. However, a real detailed study is not possible without using QFT. Radioactive decay is the simplest example of decays which are detected at macroscopic scale but they cannot be fully understood without incorporating quantum mechanics at the individual particle level and further details can only be investigated using QFT at high energies. However, radioactive decays are identified from their products or the final states and we can generally express the nuclear decays and they are written in terms of general nuclei X and Y for the atomic number as Z (number of protons) or the atomic weight A corresponding to the sum of protons and neutrons. Some of the examples of well-known decay rates are:

(i) **Alpha decay**

Processes such as:

$$X_Z^A \to \text{He}_2^4 + Y_{Z-2}^{A-4}.$$

These processes are identified as alpha decay, where a helium nucleus with two protons and two neutrons (He_2^4) is called an alpha particle.

(ii) **Beta decay**

Processes such as:

$$n \to e + p + \overline{\nu}_e$$

are an underlying process which is expressed as a generalized nuclear process which causes the production of an electron.

$$X_Z^A \rightarrow {}_{-1}e^0 + Y_{z+1}^A$$

(iii) **Gamma decay**

Photons are also called gamma particles and a general form of gamma decay can be written as:

$$X_Z^A \rightarrow Y_Z^A + \gamma$$

(iv) **Positron decay**

These decay processes are not very common but they produce a positron instead of electron. This process can be considered as a different channel of the same process and is given as:

$$X_Z^A \rightarrow {}_{+1}e^0 + Y_{z-1}^A$$

(v) **Electron capture**

Another channel of the same beta process can be identified as electron capture:

$$X_Z^A + {}_{-1}e^0 \rightarrow Y_{z+1}^A$$

where the electron capture can be considered as a scattering process. In general, all particle processes including both the decays and scattering due to a particular interaction can be written as:

$$\langle f\,|\hat{S}|\,i\rangle$$

which shows the incoming initial state $|i\rangle$ changes into the outgoing final state $<\langle f\,|$ in the presence of the interaction represented by S-matrix ($\hat{S}$) as an operator. This operator matrix $\hat{S}$ describes the interaction of incoming and outgoing states.

When there is a possibility of several incoming states and several outgoing states, then $\langle f\,|\hat{S}|\,i\rangle$ represents a matrix element for a particular process. Every process is written in terms of a matrix $\mathscr{M}$. Before we write the matrix element for a given process, we need to first understand if we can visualize how these processes take place. Generally, in scattering theory and in particular in quantum mechanics, the Born approximation takes the incident field in place of the total field as the driving field at each point in the scatterer. The Born approximation is named after Max Born who proposed this approximation in the early days of quantum theory development. It is the perturbation method applied to scattering by an extended body. It is accurate if the scattered field is small compared to the incident field on the scatterer. For example, the scattering of radio waves by a light styrofoam column can be approximated by assuming that each part of the plastic is polarized by the same electric field that would be present at that point without the column, and then calculating the scattering as a radiation integral over that polarization distribution.

It is also worth mentioning, processes that span from high energy to low energy can decay into smaller masses (from larger ones) within inelastic processes. In high-energy relativistic physics, due to the interconversion of mass and energy, high-mass particles can be produced from lower mass, increasing the number of possibilities to a large value. In this case, a complete study of decay rates may involve more than one interaction unless we separate them energetically. QED processes can take place on almost all energies, whereas for weak and strong interaction we need a much smaller scale of study and high enough energies for the relevant interactions.

8.4.2 Decay rates

The matrix element for a decay rate can be written as:

$$\mathcal{M} = \overline{u(p)} ie\gamma_{\mu} u(p) \tag{8.38}$$

$$\frac{d\Gamma}{d\theta} = \int d\theta \, |\mathcal{M}|^2 \tag{8.39}$$

θ is the scattering angle and the above equation reads that the scattering probability between the angle θ and $\theta + d\theta$ for the given decay mode is represented by the matrix $\mathcal{M}$. Integration over the angle θ will give the total decay probability for the given mode.

The lifetime of a particle can only be obtained if we can take a sum of all possible decay processes. In the calculation of the decay rate we might be interested to know the probability of a particular mode as compared to the lifetime of the particle.

8.4.3 Scattering cross-sections

The scattering from a point source is a quantum mechanical phenomenon which can only occur at quantum scale. However, it is a much more complicated phenomenon where we need to incorporate the possibility of various decay modes including the variation in target particles and discussing different modes of scattering or different channels among the same particles. The matrix element or an S-matrix for a scattering process can be written incorporating multiple initial states and summing over all possible final states. We also needed to incorporate all channels of scattering processes.

$$\mathcal{M} = \bar{u}(p_1) ie\gamma_{\mu} u(p_2) D^{\mu\nu}(k) \bar{u}(p_1{}') ie\gamma_{\nu} u(p_2{}') \tag{8.40}$$

For the solid angle Ω:

$$\frac{d\sigma}{d\Omega} = \sum_i \int dk \, |\mathcal{M}|^2 \tag{8.41}$$

Averaging over the initial state and summing over the final states is also required. An integration over the solid angle $d\Omega$ gives the total scattering cross-section. Scattering amplitude is then obtained by summation over all the initial states and averaging over final states. Some of the examples of QED elastic scattering are very well-known and their calculations provide a good exercise to understand the calculation method. These examples are Rutherford scattering ($e^-p^+ \to e^-p^+$), Compton scattering ($e^-\gamma \to e^-\gamma$), Bhabha scattering ($e^-e^+ \to e^-e^+$), Mott scattering ($e^-e^- \to e^-e^+$), etc. An interested reader can calculate them for practice as most of them are part of every introductory particle physics book.

8.4.4 Radiative corrections

At sufficiently high energies, all the particles have the ability to create and reabsorb a virtual particle which may not be detectable at all. These virtual loops are indicated as radiative corrections and could include both fermion loops and boson loops. A study of these virtual loops is made in the perturbation theory of gauge theory. Another requirement of a physically acceptable gauge theory is that it can be renormalized by taking a sum of all the perturbation diagrams of the same number of loops. Renormalization of gauge theories requires the cancellation of singularities contributed by all the loops. As of now, QED is the only perfectly renormalizable gauge theory as it has been tested that it is renormalizable even if we add a large number of loops. The reason of getting a successful perturbative series is that its coupling constant $\alpha \sim \frac{1}{137}$ and the perturbative expansion in the order of perturbation as $\alpha = \frac{e^2}{4\pi}$ is a coupling constant which gives a convergent perturbative series.

8.5 A common language of field theory

The calculations of particle processes become very complicated as they may involve several particles and these processes can go through various channels. It makes keeping account of all the information a little too complicated. Feynman in 1948 proposed a very convenient way to create a pictorial representation of particle processes in two-dimensional space, known as Feynman diagrams. Not only that, but he could also develop a general method to represent every portion of the diagram in a standard way to conveniently write the matrix element for a Feynman diagram which basically describes a particle process. His method became so popular that it is now adopted as a standard technique by all high energy physicists as a common language. It may be drawn in two ways. We can use time as a real component and space on the imaginary axis as ($\vec{x}$, t) or taking time real and space imaginary (t, $\vec{x}$). The imaginary-time presentation is called Euclidean space and the real-time space generated by Wick's rotation is called Minkowski space.

The QFT provides a framework to calculate the decay rates and evaluate the scattering amplitudes from Green's functions by computing of incoming and outgoing states and focusing on the possible interaction dynamics of the QFT. However, calculating the Green's functions for interacting fields can be complex. To simplify this, we use Feynman diagrams, which offer a visual and calculational method for evaluating these Green's functions, organized according to the structure of the Lagrangian of the theory.

8.5.1 Feynman diagrams

Feynman diagrams represent the dynamics of interacting particles. Each diagram incorporates all the incoming and outgoing states along with the mediators or the corresponding interaction. To construct Feynman diagrams, we need the Feynman rules, which are derived from the interaction Lagrangian. For an interacting scalar field theory with a typical interaction term $\mathscr{L}_{\text{int}} = -\frac{1}{4}F^{\mu\nu}F_{\mu\nu}$.

A standard method to draw Feynman diagrams in four-dimensional space–time coordinates was drawn in which an incoming state to the point of interaction is drawn from left to right and an outgoing state coming out of the same point of interaction. At the point of interaction, incoming state changes into the outgoing state with the emission or absorption of a mediator. The point of intersection of a mediator with the incoming and outgoing state is called the vertex. An incoming spinor for a particle is represented as $u(p_1)$ and outgoing fermion state as $\bar{u}(p_2)$, whereas an incoming antiparticle is represented as $v(p'_1)$ and an outgoing fermion state as $\bar{v}(p'_2)$. An incoming photon (a vector boson) is associated with the vector field A_μ such that the mediator heads to another vertex on the other end. Since the mediator emits and is absorbed and skips detection we call it a virtual particle.

Basic conservation rules of four-momentum coordinates are obeyed by these particles and a mediator can only communicate between two vertices if those conservation rules are protected. A vertex is represented by the interaction at the vertex and keeps the relevant information. The QED vertex is indicated as $ie\gamma_\mu$, whereas, for a complete process, the second vertex can then be indicated by $ie\gamma^\nu$. The matrix element for this process is indicated by applying the conservation rules including a matrix g_ν^μ such that the $g_\nu^\mu \gamma_\mu \gamma^\nu = \gamma^2 = 4$.

This transformation matrix can be represented as g_ν^μ has two different representations. The Euclidean matrix, with imaginary time, can be then be written in the form:

$$g_\nu^\mu = \begin{bmatrix} -1 & 0 & 0 & 0 \\ 0 & 1 & 0 & 0 \\ 0 & 0 & 1 & 0 \\ 0 & 0 & 0 & 1 \end{bmatrix} \tag{8.42}$$

whereas the same matrix in Minkowski space with real-time and imaginary spacial coordinates attains the form as:

$$g_\nu^\mu = \begin{bmatrix} 1 & 0 & 0 & 0 \\ 0 & -1 & 0 & 0 \\ 0 & 0 & -1 & 0 \\ 0 & 0 & 0 & -1 \end{bmatrix} \tag{8.43}$$

whereas the virtual boson propagator in momentum space is written as:

$$D_\nu^\mu = \frac{g_\nu^\mu}{k^2 - M^2} \tag{8.44}$$

where k is the momentum transfer at the vertex defined as $\vec{k} = \vec{p} - \vec{p}\,'$ and M is the rest mass of the boson, which is zero for a photon. However, the fermion propagator for the fermion momentum p is written as:

$$S_F(p) = \frac{\not{p} - m}{p_2 - m^2} \tag{8.45}$$

where p is the e-momentum of a fermion and m its mass.

8.5.2 Feynman rules for Feynman diagrams

The K–G equation and the Dirac equation give the relativistic equation of motion of a single particle. While discussing the particle processes, the individual equation of every particle is incorporated to determine the probability of a particular process. Feynman came up with a graphical representation of particle processes to develop a very convenient approach to determine the probability of occurrence of various processes using a Feynman diagram. This approach provided an excellent and manageable way to determine the probability of these processes using Feynman diagrams. We just present here the Feynman diagrams and Feynman rules to demonstrate how to use the Feynman rules to evaluate the Feynman diagrams.

These Feynman diagrams provided a convenient way to write the *S*-matrix which is an initial step to calculate the decay rates and scattering processes. Before getting into further detail, we write a general approach to drawing Feynman diagrams. A general diagram is represented by initial and final states as external lines and the exchanging particles as internal lines such that:

1. **Propagators**: Represent the probability amplitude for a particle to propagate from one point to another. Each internal line in the Feynman diagram corresponds to a propagator. The mathematical form of the propagator depends on the type of particle and the field it represents. For a scalar field of mass m, the propagator in momentum space is given by:

$$\frac{i}{p^2 - m^2 + i\varepsilon},$$

whereas the vector boson propagator is given as:

$$\frac{g_{\mu\nu}}{k^2 - m^2 + i\varepsilon}.$$

2. **Vertices**: They are the points that represent where three particles gather at the same points where actual field interactions occur. Each vertex in the Feynman diagram corresponds to a factor derived from the interaction terms in the Lagrangian. For an interaction term of the form $-\lambda\phi^n$, where λ is the coupling constant, each vertex contributes a factor of $-i\lambda$ and connects n lines.

3. **External legs**: Representing the incoming and outgoing particles, which correspond to the asymptotic states in the *S*-matrix element. Each external line contributes a wavefunction, $e^{+ip\cdot x}$ for an incoming particle and $e^{-ip\cdot x}$ for an outgoing particle, where p is the particle's momentum and x is the position. For calculations involving scattering amplitudes, the external particles are often placed on-shell, meaning their momenta satisfy the requirement of real particles $p^2 = m^2$.

To calculate a scattering amplitude $\mathcal{M}$ for a given process, follow these steps:

1. **Identify all possible diagrams:** Draw all distinct Feynman diagrams with the specified number of external particles and interaction vertices for the given order in perturbation theory.
2. **Assign momenta to each line:** Assign momenta p to each external line and assign momenta to internal lines, ensuring conservation of momentum at each vertex.

3. **Apply Feynman rules:** For each element in the diagram, apply the Feynman rules:
    - Assign a propagator factor to each internal line.
    - Assign a vertex factor to each interaction vertex.
    - Include a wavefunction for each external line, if applicable.
4. **Integrate over internal momenta:** For each independent loop in the diagram, integrate over the loop momentum with a phase factor $\int \frac{d^4q}{(2\pi)^4}$.
5. **Combine and simplify:** A matrix element $\mathscr{M}$ is comprised of all factors from propagators, vertices, and external lines.

- **Symmetry factors:** If there are identical particles or symmetries in the diagram, divide by a symmetry factor S to account for over-counting equivalent configurations.
- **Coupling constants and interaction strengths:** The coupling constants in the Lagrangian, such as λ in a ϕ^4 theory, appear as multiplicative factors in the final amplitude, representing interaction strengths.

For a typical field theory:
- Each internal line contributes a propagator.
- Each vertex contributes a factor based on the interaction term.
- Each external line is associated with an initial or final state wavefunction.
- Integrate over each independent loop momentum.

Thus, the Feynman rules provide a structured method to compute scattering amplitudes, reducing the complex dynamics of quantum fields into manageable mathematical terms

8.5.3 Perturbation in gauge theories

Feynman diagrams and Feynman rules provide a mechanism to calculate the perturbative series in gauge theories. The perturbation theory of QED (see figure 8.1), for example, redefines the basic parameters of the theory as the electron mass gets perturbative representation as:

$$m_R = m + \delta m = m(1 + a_1\alpha + a_2\alpha^2 + a_3\alpha^3 + \cdots) = \sum_{i=0} m(a_i\alpha^i)$$

$$e_R = e + \delta e = e(1 + a_1\alpha + a_2\alpha^2 + a_3\alpha^3 + \cdots) = \sum_{i=0} e(a_i\alpha^i)$$

whereas m_R and e_R correspond to the renormalized mass and charge of electron. The electron wavefunction is also renormalized using the self-mass corrections δm. The calculation of perturbative contribution becomes possible using Feynman rules and Feynman diagrams only. Now the renormalization techniques are developed for every fundamental interaction theory. Electroweak renormalization and the renormalization of QCD can be proved using special techniques such as the effective

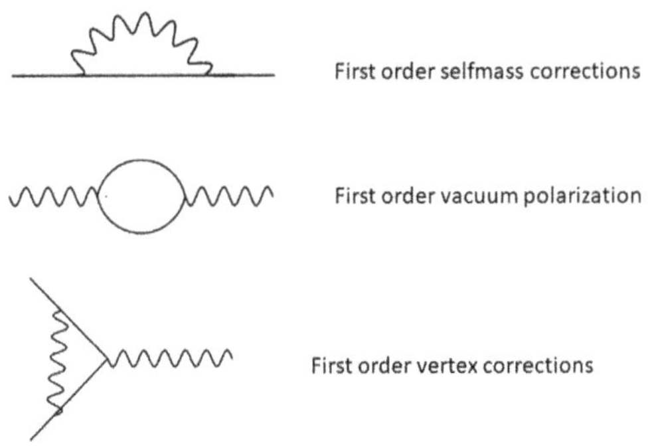

Figure 8.1. First order perturbative diagrams of QED.

potential of perturbative expansion. A brief summary of Feynman rules of QED is included in the appendix for convenience. These Feynman rules can help us write a Feynman diagram for decay or scattering processes in QED. These Feynman rules provide a mechanism to calculate the perturbative calculations and prove the renormalizability of gauge theories.

IOP Publishing

Conceptual Approach to Quantum Electrodynamics

Samina S Masood

Chapter 9

Applications of quantum electrodynamics

9.1 Introduction

Quantum electrodynamics (QED) is the only known interaction theory which is equally applicable from quantum mechanical scale to cosmic scale. It studies individual particle behavior and integrates it into the large-scale description of classical electrodynamics using proper approximations. It is therefore possible to consolidate the three-dimensional classical electrodynamics and four-dimensional QED to cover the entire range of coordinate space. QED is therefore considered to be the most successful fundamental interaction theory which establishes a correlation between principles of QED and classical electrodynamics (just like the correlation between classical physics and quantum mechanics or the correspondence between relativity and non-relativistic classical physics). QED can therefore be considered to be a single working theory of interaction between charges all the way from the independent individual particle level to the macroscopic world. Its classical forms provide a working framework for the same system at different energies. Electromagnetism works perfectly at the lowest energies and classical electrodynamics provides all the needed information about the dynamics of charges in the macroscopic world, whereas QED is needed to integrate the individual particle behavior into the many-body system.

QED is developed for electromagnetic interactions using relativistic quantum mechanics and the calculational techniques of quantum field theory (QFT) at the individual particle level. A detailed investigation of the properties of materials is performed by integrating the behavior of individual particles using the relevant physical conditions. These techniques provide resources to reveal the hidden secrets of matter in given conditions. They provide very effective tools to study the material properties in different phases. Additionally, this leads to significant improvement in technology and opens new directions of technical usage of material to develop it significantly. The application of materials research opens new venues of innovative research to reveal the hidden secrets of life and understand the universe. Its possible

doi:10.1088/978-0-7503-6054-8ch9

impact in space technology and biomedical physics cannot be ignored either. It increases the scope of technology and leads to new directions of research and development in various disciplines in science.

Material properties play a fundamental role in the development of technology. These materials can be identified by their physical, chemical or electromagnetic properties. Most of the material properties at the macroscopic level are experimentally measured and then used in the development of technology. Engineering development significantly depends on the choice of proper materials while scientists continue to explore the properties by digging deeper and deeper into materials using theoretical and experimental resources even at the subatomic and nuclear level. The mathematical tools become much more efficient in quantum mechanics where atomic and molecular behavior can be related to atomic structure and chemical processes can be understood in terms of movement of electrons incorporating their spin in the framework of quantum mechanics.

However, for the rapidly-moving electrons and nuclear matter at higher energies, relativistic effects are not negligible anymore and we have to incorporate QFT to study the correct dynamics of matter. Mathematical tools of QFT in a four-dimensional relativistic space are much more efficient and have the ability to dig deeper and deeper into material formation and study the structural impact. For this purpose, we can apply QED in its many-body formalism to incorporate the interaction among the particles at high energies. This appears as a modification to QED in many-body systems incorporating the interaction of the propagating particles through the interacting medium.

The wave–particle duality of individual particle waves and the spin statistics of quantum mechanics are now used extensively to determine the physical, thermal and electromagnetic properties of materials. So quantum mechanics is now incorporated and its utility has been established very well. However, QED was never considered to be important until its relevance for technology was established. However, it is now understood that the many-body theories need to use the raising and lowering operators and the second quantization helps to develop an oscillator model for a many-body system. These raising and lowering operators for bosons are defined as the operators associated with the increase or decrease in energy corresponding to the energy level of the system. These raising and lowering operators correspond to transitions between spin states for fermions.

Investigation of material properties in reference to astrophysics and cosmology is extensively studied using QED or even other high-energy theories of QCD and the standard model which are clearly out of scope of this book. This approach is relatively new and still in the developmental stage. However, the importance of development of technology for space exploration cannot be ignored at all. Most of the existing technology is now based on the quantum mechanical description of a material which provides methods to understand the electromagnetic properties of materials including spin and the interaction of spin with its surroundings. However, for the moment, we bypass this discussion and first investigate the relevance of QED with materials and dig deeper into material properties.

9.2 QED and materials

Classical electrodynamics gives a classical description of the behavior of electromagnetic interaction of charges, individually or collectively. Classical description of charge is related to the overall charge on an object (individual particle or a group of particles) as long as the collective net charge inside a large charged object is exhibiting a non-relativistic motion and contained inside without crossing the boundaries of a system which is unable to interact independently. The simplest form of a QED Lagrangian is then given as:

$$L_{\text{QED}} = \iota\bar{\psi}(\gamma^{\mu}\partial_{\mu} - m)\psi - \frac{1}{4}F_{\mu\nu}F^{\mu\nu} \tag{9.1}$$

QED is now accepted as a fully established gauge theory of charged particles from the individual particle scale to the macroscopic level. It is therefore possible to integrate individual particle behavior distributed at various energy levels to give a detailed description of electromagnetically interacting systems. The collective behavior of materials is determined from the individual particle behavior to discover some of the relatively unexpected features of materials. In this way, it helps to unfold some hidden properties of materials contributing to the development of technology in a more efficient way. We cannot provide a comprehensive list of all the possible applications of QED in technology, astrophysics and cosmology or summarize the usage of materials for energy storage, environment protection and proper usage of materials to develop medical science to protect human life. We only include a qualitative summary of possible applications. It is very hard to cover each and every possibility because materials science is a multi-disciplinary field of study in itself with numerous applications in technology. However, its various features are being explored in physics, chemistry and engineering directly. The possible applications to food science, space technology, medical science, biotechnology, and in almost all disciplines of engineering is being explored regularly.

QED uses a framework of second quantization which has the ability to accommodate the many-body form of electrodynamics without losing the properties of individual particles. For this purpose, the many-body Hamiltonian in quantum mechanics is described as an oscillator with the lowest energy:

$$H = H_0 + \sum_{n=0}^{n}(n + \frac{1}{2})\hbar\omega \tag{9.2}$$

where $n = 0$ corresponds to the ground state energy of the system which helps to investigate the detailed structure and behavior of matter at the individual particle level. Larger energy sources are required to look deeper into matter at the subatomic level. A detailed study of the material requires the interaction of matter with radiation and QED provides the most efficient theory to describe the interaction of radiation with matter. A lot of current development of technology has led to sophisticated technologies such as the storage of energy, quantum computing or development of low-cost lasers. Various models are also developed to investigate the details of various properties of materials. It is very hard to put a complete list of all many-body QED models. A different Hamiltonian or various potentials are used to theoretically describe various properties of systems which cannot be ignored at any level. We just present one model as an example to give an idea that QED provides a framework for the study of materials. It provides resources for *ab initio* calculation of the properties of materials. This detailed information will potentially help to design more effective structures as well. We are considering here the term 'material' for the hard materials or crystals for now. Soft materials or other condensed matter phases will be discussed later.

9.2.1 Materials and technology

Hard matter is made up of atoms which have electrons revolving around nuclei. These atoms are contributing to the formation of materials including nano-structures, solid rocks or powders. Massive objects made up of nanoparticles, crystalline structures or a composite material with very heavy mass may be sitting at rest in the lab frame or anywhere else at macroscopic level. However, they are made up of atoms and molecules that have at least mechanical vibrations and these vibrations which may not be necessarily negligible and may change their physical properties. However, we extend the study of materials to condensed matter physics to include interacting fluids or generally many-body systems which are somehow confined in a given volume. Fluid is a form of condensed matter with significant relative motion among the constituent particles with respect to one another and stays confined in a fixed volume due to mutual interaction among its constituents. In liquids or well-known fluids or plasmas equations of motion and the mutual interaction is limited to gravity. We generalize these fluid systems to small scales and extremely high energies. Electrons inside atoms are moving fast and may acquire higher energies at higher temperatures and be treated relativistically to find significant corrections. We can extend fluid mechanics for interacting fluids to quantum statistical field theories where the dominant interaction is QED which takes over gravity.

While talking about materials or crystals even at lower energies, we have to consider the vibrational motion of particles. This motion is slow but still it produces some energy. The quanta of this energy are identified as phonons, which have properties like photons. Basically, we find electrons and phonons in solid materials. Phonons can be considered as discrete units of energy associated with the vibrational mechanical energy. They are emitted and absorbed internally. For a thin metal sheet, excessive energy of electrons can be created in the form of plasmons which is related to plasma oscillation of electrons where we consider electron–phonon plasma. Surface plasma resonance (SPR) provides suitable techniques, which pushes the limits for detection of sensitive interaction of biomolecules and leads to development of low-cost and high-flexibility materials for biosensors. All these material descriptions involve quantum mechanics and the concept of second quantization for high energies.

Condensed matter physics at non-relativistic temperatures is dealt with using statistical mechanics and thermodynamics incorporating gravity, as long as the solids and fluid particles stay in non-relativistic motion among themselves. Study of the phase transition which uses the Ising model to understand the formation of crystals using the two-dimensional lattice type of structures created by the interaction of spin to create bulk materials. The Ising model is a two-dimensional model which is generalized to the three-dimensional Potts model, which is a generalized form of Ising model which incorporates gravity. The role of second quantization cannot be ignored for that purpose either. This four-states Potts model is sometimes called the Ashkin–Teller model, which was considered to be equivalent to the Potts model as well. In addition to the conventional models which are used to study the phase transition, there are other generalized models such as the Heisenberg model and the flux tube model which helped to describe confinement in QCD as well. In addition, QCD uses lattice gauge theories. This is how condensed matter physics can sometimes provide a framework for strongly-interacting particles, and a connection between high-energy physics, materials science and technology is established. Various approaches are used to apply to every physical system.

A detailed study of materials at high energies needs QED and many-body high-energy systems need to use the concepts of condensed matter physics. These theoretical models are then incorporated into computer simulations to understand the properties of matter for fixed structures in condensed matter. These models use non-relativistic quantum mechanics to incorporate spin in material study, especially for hard matter. These models are used in force fields in simulation programs based on density functional theories in various environments. These model-dependent force fields could incorporate many different types of interaction terms or include specific model-dependent terms using the Ising model or the Potts model. However, at higher energies or higher temperatures, relativistic effects have to be incorporated.

Some of the examples of QED models in materials science include the density functional theory extensively used to calculate the electronic structure of materials using the Hamiltonian incorporating second quantization in the force fields. It helps to develop computer programs for materials study including some of the combinations of very well-known programs such as Nanoscale Molecular Dynamics (NAMD), Gaussian, QuantumATK, Amber and various other programs used in physical chemistry or materials science to study particular features of polymers, nanomaterials and proteins or other biomolecular structures and the processes involved in molecular dynamics. The study of quantum materials extensively uses QED to understand the behavior of exotic materials such as topological insulators, superconductors and superfluids. The most important part of application of QED comes through the understanding of radiation with matter. This quantum mechanical model has to be extended to QED to understand the charge particle behavior incorporating spin effect as well as the relativistic energy of electrons. This interaction provides information to study quantum optics and lasers.

The perturbation theory and some of the structural details at high energies involve QED to combine classical and quantum physics and QED is used for this purpose as well. The most important application of QED is found in studying the interaction of radiation with matter which is the basis for the development of materials for very important applications including laser industry and energy production and storage techniques. Of course, it also extends to studying spatial objects including Solar System debris, and other cosmic bodies. Its applications to understand geological and environmental impacts are also being explored. Just as an example, we describe an extensively used model, called the Jaynes–Cummings (JC) model to demonstrate how various approximations are used for various models. The correct choice of the model will therefore depend on the purpose of calculations.

9.2.2 Jaynes–Cummings model

QED deals with particles which are not necessarily in bound states. To incorporate the spin statistics, the lowering and raising operators of quantum mechanics are needed to define the second quantization. The spin statistics play a big role in describing many-particle dynamics of unbound states which are distributed based on energies and follow spin statistics. One of the commonly used models in condensed matter physics is called the Jaynes–Cummings model. Starting with the definition of a number of particles in a spin state as:

$$\hat{\sigma}_z = |e\rangle\langle e| - |g\rangle\langle g|$$

where $|g\rangle$ corresponds to the ground state and $|e\rangle$ to the excited state of atoms or bosons. It gives the total Hamiltonian as:

$$\begin{aligned} H_{\text{total}} &= \hbar\omega_c + \hbar\omega_a + H_{\text{int}} \\ H_0 &= H_{\text{field}} + H_{\text{atom}} \end{aligned}$$

where evaluating the number operator N_j:

$$N_j = a_j^\dagger a_j \tag{9.3}$$

we can find the number of particles in a state j. However, considering the oscillator model, the frequency of oscillation is determined in terms of energies of two consecutive states i and j such that we can write the Hamiltonian of the system and the interaction Hamiltonian for the jth particle is given by:

$$H_j = \hbar\omega_j\left(N_j + \frac{1}{2}\right)I + \frac{1}{2}E_j\sigma_{zj} + \hbar\kappa_j(a_j^\dagger\sigma_{-j} + a_j\sigma_{+j}). \tag{9.4}$$

In the Jaynes–Cummings model, we consider a model system consisting of two identical atoms in two separate cavities encountering a photon exiting a beam-splitter, which allows non-zero probabilities for the photon to enter each cavity. We consider both atoms in their ground states initially. The Jaynes–Cummings model uses the rotating-wave approximation which describes the coupling of each atom with the radiation field. This rotating-wave approximation makes it more relevant for atoms where electrons are revolving around the nucleus. In this two-atoms model, we can even compute the atom–atom entanglement using von Neumann entropy S in terms of ρ the density of states as a measure and is expressed as:

$$S = Tr(\rho ln\rho)$$

We can also consider a case in which the two atoms–two photons systems have identical properties, but allow time-independent entanglement for non-resonant conditions in the sum of the atom–atom and photon–atom mode. Detuning is one of the methods used to decrease the strength of the largest entanglement and shorten the time of entanglement. The results support the fact that the state of the photons after emergence from the beam-splitter is entangled which does not let us treat the entangled state as a single particle. The Hamiltonian of the Jaynes–Cummings model in the rotating-wave approximation corresponds to an atom (as $j = A, B$) and its interaction with mode j such that:

$$H_{ij} = \hbar\omega_j\left(a_j^\dagger aj + \frac{1}{2}\right) + \frac{1}{2}E_j\sigma_{zj} + \hbar g_j(a_j^\dagger\sigma_{-j} + a_j\sigma_{+j}).$$

In this Hamiltonian, the first term on the right-hand side corresponds to the unperturbed Hamiltonian for particles in state j. As mentioned previously, $a^\dagger$ and a correspond to creation and destruction operator for atoms in ground state or in the excited states. Here $\sigma^\dagger_{+j=} = \sigma_{xj} + \sigma_{yj}$ and $\sigma_{+j=} = \sigma_{xj} - \sigma_{yj}$. The total wavefunction of the state can be written as:

$$|\psi(t = 0)\rangle = c_1|01\rangle + c_2|01\rangle$$

which gives the probability of atoms going from a ground state $|e\rangle \equiv |0\rangle$ to the excited state $|e\rangle \equiv |1\rangle$.

The time evolution of a state can be considered as a two-atom system with atoms in state $A = |0\rangle$ and atoms in state $B = |1\rangle$. In this case, an extension of this model is used which is called the double Jaynes–Cumming model. Usually, we can write a master equation for a cavity such that, for the given density ρ_j in state j, the time evolution can be determined by using the master equation, e.g.:

$$\frac{d\rho_j}{dt} = -\frac{i}{\hbar}[H_j, \rho_j] + \gamma[a_j\rho_j a_j^\dagger - (1/2)a_j^\dagger a_j\rho_j - (1/2)\rho_j a_j^\dagger a_j]$$

Moreover, the choice of a right material for each and every machine, equipment, tool or even simple container, etc depends on material properties and the industrial usage. It also helps in material designing and its durability depending on its shape, size and intended use of a container. Nanomaterials may now be added to increase the usage and keep the cost affordable. Detailed information of the properties of materials is very important to improve the performance and opens doors to increase the utility of materials. Detailed information about the properties of liquids or fluids helps to figure out the long-term properties of materials and the effect of their presence on the other materials as well. The properties and behavior of gases at unusually high temperatures and in unusual cosmological conditions is determined to understand cosmological objects.

9.2.3 QED in condensed matter physics

A detailed study of chemical processes is also possible if second quantization is incorporated to study interacting fluids. This second quantization allows us to incorporate the spin statistics along with the quantum mechanical description of electrodynamics. However, the interaction of radiation with matter and a study of secondary processes is made possible using QED formalism which can be generally applied from fundamental particles to all the way to individual charged particles. QED is the most generalized description of electrodynamics which incorporates relativity, quantum mechanics and statistical mechanics in a single theory. We identify this framework as a quantum statistical field theory which brings all the mathematical tools together.

QED can then be equally applied to relativistic electrons and non-relativistic ions or protons and is successfully applied to particle processes. QED is an abelian gauge theory as we learn its nature from classical electrodynamics. It follows a particular symmetry of the (magnitude of) electromagnetic interaction and the abelian nature is typically related to the fact that two interacting particles feel the same amount of force at the same distance. However, it lays down a foundation for the development of gauge theories in terms of the Lagrangian formalism. Its generalization to non-abelian gauge theory demands a non-abelian group representation of the symmetries of a Lagrangian. However, we accept the gauge invariance as a basic requirement of a physical theory. Gauge theories and the group representations together provide basic tools to describe the dynamics of fundamental particles. Non-abelian theories are not the topic of this book so we will just concentrate on QED here.

Cosmological description of the universe requires the space–time coordinates to describe the fundamental contents of the universe in the form of mass and energy. Relevance of QED in particle physics, astrophysics and cosmology is relatively more obvious and can be discussed in detail.

9.3 Quantum statistical field theories

Quantum field theories can be extended to many-body systems to apply fundamental interaction theories at relativistic energies. However, QED has more physical applications to use many-body approaches in a quantum statistical background. Many-body QED is extensively used in materials science and technology along with its applications to technology. Its implications to develop remote techniques to explore deeper into astrophysics and cosmology cannot be ignored either. However, due to the larger number of unknown parameters of the theory, the development of exact mathematical expressions becomes much more complex and computational tools are developed to help out the calculations.

On the other hand, specific mathematical tools are used in the quantum statistical theories to solve complex many-body equations. Specific mathematical techniques such as functional methods and path-integral techniques are used and even special modified forms of mathematical techniques are developed to solve complex statistical problems. Quantum statistical field theories are used incorporating the distributions of energies using spin statistics in four-dimensional space of multi-particle systems.

Quantum statistical theories of fundamental interactions cannot be adopted for physical applications unless they pass the test of gauge invariance and are proved to be renormalizable as well. The QED Lagrangian can successfully pass the gauge invariance requirement for particular gauges such as Feynman gauge, Lorentz gauge and Coulomb gauge. Gauge theory is in itself an interesting subject and specific gauges seem to be more relevant for specific problems. QED is therefore considered to be a very interesting subject for practical applications in applied science and technology as well. Skipping all the complex technical details, we can give a brief overview of the statistical formalism used to develop this approach and include a few relevant references at the end of the book for further reading for interested readers. It is also worth mentioning that this theory is still in the developmental stage and a comprehensive book on the topic with all the updated information may not be available in a book form yet.

Many-body (statistical) formulation of QED in an interacting medium with charged particles and electromagnetic radiation background incorporates the statistical parameters like temperature, density and phase space along with the defining parameters of the interaction theory of electromagnetism. Thermal variation in space is related to the variation in the Euclidean matrix. The study of particle propagation in Euclidean space is identified as imaginary-time formalism (or Matsubara formalism). Addition of probabilities of interaction of the propagating particles seems to give an option to examine every individual state and the properties which a system acquires going through a dynamical process. However, a detailed study of dynamics of every intermediate state is not practically possible for a highly-energetic system in the presence of various unknown parameters. The same treatment can be used for bosons and fermions using Bose–Einstein and Fermi–Dirac distribution functions. These techniques are developed in Euclidean space or imaginary-time formalism and then transformed to Minkowski space, giving a framework for real-time formalism through Wick's rotation. Distinguishing features of both formalisms and a brief comparison of both models are given here. Thermal equilibrium is always an underlying assumption.

9.3.1 Imaginary-time formalism

The development of statistical QFT started from the statistical representation of Green's function in coordinate space in a statistical system as:

$$G_\beta(x_1, \ldots, x_j) = \frac{Tr\ e^{-\beta H} T\ \phi(x_1) \cdots \phi(x_j)}{Tr\ e^{-\beta H}}, \tag{9.5}$$

where ($\beta = \frac{1}{k_B T}$) is the reciprocal of temperature T, and Boltzmann constant $k_B = 1$ giving $\beta = 1/T$. H is defined as the Hamiltonian for the self-interacting field $\phi(x)$ and its existence is incorporated in any relevant fundamental interaction theory. The finite-temperature effective action, $\Gamma(\overline{\phi})$ is a generating functional for a single-particle field expressed as:

$$\Gamma(\overline{\phi}) = W^\beta(J) - \int d^4x \phi(x) J(x). \tag{9.6}$$

which gives an irreducible Green's function and is defined by the following equations such that:

$$\overline{\phi}(x) = \frac{\delta W^{\beta}(J)}{\delta J(x)},$$

where:

$$W^{\beta}(J) = -i \ln Z^{\beta}$$

and the partition function z attains the form:

$$Z^{\beta}(J) = Tre^{-\beta H} T \exp\left[i \int d^4x \phi(x) J(x) \right] Tre^{-\beta H}, \tag{9.7}$$

Now, the $J(x)$ dependence can be eliminated by expressing $J(x)$ in terms of the effective potential $\overline{\phi}(x)$ such that:

$$\frac{\delta \Gamma(\overline{\phi})}{\delta \overline{\phi}(x)} = J(x), \tag{9.8}$$

where $\overline{\phi}(x)$, evaluated at $J(x) = 0$, gives the thermodynamic average of the field $\phi(x)$ such that:

$$\overline{\phi}(x)|_{J=0} = \frac{Tre^{-\beta H} \phi(x)}{Tre^{-\beta H}}. \tag{9.9}$$

The generating functional is helpful in studying the spontaneous symmetry breaking. It was assumed that a symmetry is possessed by the Hamiltonian of the system under normal conditions, such that we get $\overline{\phi} = 0$ at $J = 0$. Alternatively, the symmetry breaking is implied for $\overline{\phi} = \not{0}$ and:

$$\frac{\delta \Gamma^{\beta}(\overline{\phi})}{\delta \overline{\phi}(x)} = 0.$$

This shows that the above equation should be independent of x and we do not expect spontaneous violation of translational invariance. Therefore, the function $\Gamma^{\beta}(\overline{\phi})$ can be studied for constant $\overline{\phi}$. The finite-temperature effective action $\Gamma(\overline{\phi})$ is a generating functional for a single-particle field, and an irreducible Green's function can be defined by the following equations. When the generating potential reaches its unique minimum value:

$$\frac{\delta \Gamma^{\beta}_{eff}(\overline{\phi})}{\delta \overline{\phi}(x)} = 0. \tag{9.10}$$

This effective potential can be used in the equation of motion and replace the non-interacting potential. A single-particle irreducible Green's function is then generated for a particle at rest. This helps to write the statistical propagators for fermions and bosons which can depend on statistical parameters. The periodic boundary conditions are imposed on the statistical equation of motion which can be derived for the two-point Green's functions. The free scalar field in the presence of the heat bath is known to satisfy:

$$(\Box + m^2)D_\beta(x - y) = -i\delta^4(x - y),$$

where the two-point temperature-dependent function of spinless fields $D_\beta(x - y)$ is obtained as a solution of this equation from the corresponding Green's function equation as:

$$D_\beta(x - y) = \frac{Tre^{-\beta H}T\phi(x)\phi(y)}{Tre^{-\beta H}} = \langle T\phi(x)\phi(y)\rangle, \tag{9.11}$$

The boundary conditions are required to solve the above equation in coordinate space. We use imaginary time such that the time argument for D_β is continued to stay in the imaginary-time interval $0 \leq -i(x - y)_0 \leq -i\beta$ and the time ordering for the imaginary time is defined, for $ix_0 > iy_0$, as:

$$\langle T\phi(x)\phi(y)\rangle = \langle \phi(x)\phi(y)\rangle => D_\beta(x - y) \tag{9.12}$$

and for $ix_0 < iy_0$:

$$\phi(x)\phi(y) > = D_\beta^<(x - y) \tag{9.13}$$

and the imaginary-time interval is given as $|0, -i\beta|$ and the corresponding state is given by:

$$e^{i(-i\beta)H}\ \phi(0, \vec{x})e^{-i(-i\beta)H} = \phi(-i\beta, \vec{x}).$$

such that:

$$(Tr\ e^{-\beta H})D_\beta^<(x - y)\Big|_{x_0 = 0} = Tr\ e^{-\beta H}\quad {}^>D_\beta(x - y)\Big|_{x_0 = -i\beta}$$

Now, using the periodic boundary conditions, we can obtain:

$$D_\beta(x - y)\Big|_{x_0 = 0} = D_\beta(x - y)\Big|_{x_0 = -i\beta}.$$

In the imaginary-time domain, D_β may be represented by Fourier series:

$$D_\beta(x - y) = \frac{1}{-i\beta}\sum_n e^{-i\omega_n x_0}\int\frac{d^3p}{(2\pi)^3}\frac{1}{(-i\beta)}\sum_n e^{i\omega_{n'}x_0}\int\frac{d^3p'}{(2\pi)^3}D_\beta(\omega_n, \vec{p}, \omega_{n'}, \vec{p}\,'), \tag{1.22}$$

where:

$$D_\beta(\omega_n, p, \omega_{n'}, p') = -i\beta\delta_{nn'}(2\pi)^3\delta^3(\vec{p} - \vec{p}\,')D_\beta(\omega_n, \vec{p}),$$

which in its diagonalized form gives the Matsubara frequencies (energies) as:

$$\omega_n = \frac{2\pi n}{-i\beta},$$

for $n = 0 \pm 1$. Hence:

$$>D_\beta(x - y)\Big|_{x_0 = 0} = D_\beta(x - y)\Big|_{x_0 = -i\beta},$$

for:

$$D_{\beta}^{<}(x-y)\Big|_{x_0=0} = D_{\beta}(x-y)\Big|_{x_0=0},$$

gives:

$$D_{\beta}(x-y)\Big|_{x_0=-i\beta} =^{>} D_{\beta}(x-y)\Big|_{x_0=-i\beta}.$$

Now, combining the above equations, we obtain:

$$D_{\beta}(x-y)\Big|_{x_0=0} = D_{\beta}(x-y)\Big|_{x_0=-i\beta},$$

which indicates that the required boundary conditions are imposed on the corresponding solution.

For the development of many-body field theory in Euclidean space (or imaginary-time formalism), we start from the partition function:

$$\mathscr{Z} = Tr(e^{\beta\mathscr{H}}) = \prod_r e^{-}(\beta E) \tag{9.14}$$

where $\beta = \frac{1}{k_B T}$ for k_B the Boltzmann constant which is set to be equal to 1 and T is the temperature and $E = \sum_n E_n$ such that:

$$\mathscr{H} = H_0 + H'.$$

Then the Hamiltonian density from Dirac equation is written as:

$$\mathscr{H}(x,t) = \psi^{\dagger}(x,t)(\not{p} - m)\psi(x,t) \tag{9.15}$$

where $\mathscr{H}'$ is the interaction Hamiltonian density expressed as the sum over n states in the physical system and contributes to the energies of the particle such that the frequencies of the propagating particles are given by Matsubara formalism. Particle propagators are then modified and are contributed to the four-momentum of particles as $\omega = \omega_0 + \omega'$ such that $\omega' = \sum_n \frac{E_n'}{\hbar}$. The modification in frequencies due to the thermal background ω' keeps an account of the interaction of propagating particles with a thermal medium and are called Matsubara frequencies. These modified frequencies contribute to the particle propagators through their four-momenta. Spin statistics is used to study the distribution of energy and contribute to phase space.

This summation in energies corresponds to the summation of the interaction potential of particles with each and every intermediate state. The availability of intermediate potential depends on the fact of how many intermediate states are interacted with the propagating particles. Keeping an account of all intermediate particles is tricky so in the Hamiltonian, one can simply add an effective potential instead based on all the interacting frequencies. This is how we define the effective potential. The effective potential is constructed in a way that the physically acceptable theory can be formulated by the proper choice of the effective potential. This choice of the effective potential makes this theory a model-dependent theory based on the choice of the relevant potential. These potentials may need modifications with the change of statistical conditions.

The imaginary-time formalism apparently has the ability to track all the intermediate steps showing the dynamics of the development of the system in detail. However, this summation is not possible as all the intermediate states are not even known. So the path-integral methods are used as mathematical techniques to solve these problems. The choice of the effective potential and the path-integral techniques hide a lot of necessary details of intermediate states and the theory and several competitive forms of effective potentials are used to solve physical problems in an acceptable way. However, this approach introduces some shortcomings in the theory as well.

First of all, thermal effects cannot be written in terms of four-dimensional parameters at the individual particle level. Therefore, the statistical background just affects the relative momentum which is changed due to the particle energies and the interaction with other particles. In this situation, the gauge invariance is broken as the entire integrand cannot show the variation in four-momenta at the individual particle level. The relative momentum and energy of the system is then expressed in terms of the effective potential only. We can generally give a list of some of the major shortcomings associated with the imaginary-time formalism.

1. Gauge invariance is broken in this formalism because single-particle momentum is not affected by temperature, whereas total energy is affected due to the change in relative motion. However, the Lorentz invariance is established at the individual particle level.

2. Using the effective potential approach, the order-by-order cancellation of singularities cannot be proved, which is one of the basic requirements of a perfect gauge theory. However, a suitable potential can be found where some approximations can be applied to find an acceptable way to renormalize the theory, if nothing else works. The validity of QED from microscopic to macroscopic scales should be proved by showing the perfect renormalizability of the theory at all energies.

3. Effective potential is used to calculate the effective parameters of the theory to a good approximation. However, a positive feature is that the choice of appropriate potential at high energies can take an overall effect without tedious calculation at each level of interaction separately.

Matsubara formalism works very well for perturbative QCD as the renormalization is already a known issue in QCD. Quarks can participate in electroweak and strong interactions simultaneously. A large coupling constant of QCD makes it difficult to handle the renormalizability of the theory. In addition, several unknown parameters may be required to understand rather strange behavior of QCD whose sources are still not very clear. A good approximate potential gives a working scheme of calculation to acceptable approximation within our limitations and a more successful computational model provides the workable scheme.

9.3.2 Real-time formalism

A calculational scheme of finite-temperature field theory was still needed to check the validity of the theory and understand the validity of various calculational schemes. For this purpose, Minkowski space or the real-time formalism is a more appropriate approach. This approach helped to re-establish gauge invariance at the cost of Lorentz invariance by choosing the rest frame of the heat bath, through which the propagation of particles is studied. This is a totally acceptable physical condition because it just gives the calculation of a system at the given instant of time when thermal equilibrium can be safely applied.

The Fourier series can then be converted into integral form by using continuation of energies from the corresponding discrete energies of imaginary-time formalism. In other words, the overall change in energy is considered instead of looking at each individual particle's energy. In this way, it can hide some details about the dynamical path but the overall changes can be calculated. One defines the density of states as:

$$\rho(k) = D_\beta(k_0 + i\varepsilon, \overrightarrow{k}) - D_\beta(k_0 - i\varepsilon, \overrightarrow{k}).$$

and ignore the intermediate stages by evaluating integrals in between the boundaries of the entire parameter space.

$$D_\beta(\omega_n, \overrightarrow{k}) = \int_0^{-i\beta} dx_0 e^{-(2n\pi\beta)x_0} \int d^3x e^{-i\overrightarrow{k}\dot{\overrightarrow{x}}} D_\beta(x),$$

In this case, the Fourier transform of Green's function in momentum space is expressed as:

$$D_\beta(p) = \frac{1}{p^2 - m^2}, \tag{9.16}$$

and the density of states can then be represented as:

$$\rho(k) = 2\pi\varepsilon(k_0)\delta(k^2 - m^2),$$

which gives the probability of interaction with real particles indicated by $\delta(k^2 - m^2)$ for all the particles within the entire momentum space which allows us to represent the boson propagator as:

$$\overline{D}_\beta(p) = \frac{i}{k^2 - m^2 + i\varepsilon} + \frac{2\pi}{e^{\beta E} - 1}\delta(k^2 - m^2), \tag{9.17}$$

with:

$$E = \sqrt{\overrightarrow{k}^2 - m^2}.$$

A similar analysis can be done for fermions to obtain the fermion Green's function as:

$$\overline{S}_\beta(p) = \frac{i}{\not{p} - m + i\varepsilon} - \frac{2\pi(\not{p} + m)}{e^{\beta E_p} + 1}\delta(p^2 - m^2). \tag{9.18}$$

The fermion and boson propagators in the real-time formalism are defined by the propagators with bars given as $\overline{D}^\beta(p)$ and $\overline{S}_F^\beta(p)$ which correspond to boson and fermion propagators, respectively. These propagators are used in conjunction with ordinary Feynman rules to calculate Feynman diagrams. Alternatively, one may use the functional techniques to calculate a Feynman amplitude in finite-temperature perturbation theory of any order of perturbation. It is one of the accepted facts that at higher energies where energy–momentum (momentum–space) becomes more convenient as compared to the space–time coordinate (coordinate–space). Fourier transform is the most convenient mathematical relation to transform everything from coordinate space to momentum space and convert it back using the inverse Fourier transform. However, Minkowski space is definitely preferable for thermal study as time and energy are both directly measurable (observable) in Minkowski space as compared to the real position and momentum of relativistic particles in Euclidean space. Gauge invariance is re-established in this theory by the choice of a preferred frame of reference. Since the integration is performed over all the phase space, a choice of the preferred frame of reference can be made. In the real-time formalism, all the calculations can be done in the rest frame of a hypothetical heat bath. This heat bath is conceptually equivalent to a phase space which is created by the entire range of four-momenta and the four-coordinate space. Taking the rest frame of the heat bath we reincorporate the gauge invariance at the cost of Lorentz invariance. And the breaking of Lorentz invariance is acceptable for a physical system. Therefore, some of the salient features of the real-time formalism are worth mentioning and we briefly summarize them here to be able to compare both formalisms.

1. Gauge invariance of the theory is re-established at the expense of Lorentz invariance. All the calculations are done in the rest frame of the heat bath. However, the scale of the heat bath is not limited. It just corresponds to the entire phase space associated with the system.

2. Using the effective potential approach, the order-by-order cancellation of singularities cannot be proved, whereas it can be shown easily using the real-time formalism. The most interesting feature of the real-time formalism is that the temperature-independent part of the propagator appears isolated from a temperature-independent regular term, especially at the first loop level. So we can simply evaluate the additive contribution which is clearly seen in equations (9.17) and (9.18).

3. Renormalization constants are calculated individually at any loop level. Order-by-order cancellation of singularities can be checked at every loop or those conditions are found where the singularities are removed to find the physically possible conditions. Thermal contributions can be compared with temperature-independent contributions at various loop levels.

The real-time formalism is the only formalisms to understand the renormalizability and the renormalization constants give the physically measurable values of the parameters of the theory and can substitute the effective parameters.

Therefore, both real-time and imaginary-time formalism have their own pluses and minuses. Moreover, the perfect matching of results obtained from both calculations is not always as found during the integration in real-time or summation in imaginary-time formalism: some underlying approximations are involved. Also, the Lorentz invariance and gauge invariance may affect the results as well. One of the drawbacks of real-time formalism is that due the breaking of Lorentz invariance, the order of integration between energy and momentum do not commute. However, the detailed comparison or the working knowledge of both of these approaches is out of the scope of this book. In addition, finite-temperature formalism is a relatively new field and it is still developing with the discovery of cosmic objects where its applications are unavoidable. Therefore, a detailed study of the current literature is required to understand the theory in detail with its applications.

9.3.3 Perturbation theory in QED

Real-time formalism plays a very effective role in QED. It becomes more relevant because it proves the renormalization of QED for physically acceptable ranges of statistical parameters providing very effective tools to understand the physics of many-body systems beyond the collective behavior. A renormalization scheme of QED provides a framework to calculate all the effective parameters of QED in context with the hot and dense media. The renormalization constants of QED depend on the temperature and chemical potential of electrons (or other relevant charged particles) which are in thermal equilibrium with the heat bath. The interaction of radiation with charged particles in the physical systems is the basic source of energy in the heat bath and is used extensively in the early universe and superdense stars due to relevant ranges of temperatures and densities. In an ultra-relativistic electron–photon system, termed the QED system, the renormalization constants of QED computed over different temperature and chemical potential ranges are considered as the physically measurable parameters of the theory.

For the purpose of renormalization of QED, standard Feynman rules are employed by incorporating the Bose–Einstein distribution for bosons and the Fermi–Dirac distribution for fermions. This substitution allows for the incorporation of temperature and density effects from the quantum statistical background of the interacting fluid. The QED Lagrangian, using the typical field theory notation, is then formulated with the renormalization constants of QED denoted by δm or Z_1, Z_2 and Z_3 called mass, wavefunction and charge renormalization constants, typically calculated for electrons considering them the lightest form of charged matter that exhibits relativistic motion easily. A complete Lagrangian of QED with the renormalization constants can then be written as:

$$\begin{aligned} L = &-\frac{1}{4}F^{\mu\nu}F_{\mu\nu} - \overline{\psi}[\gamma_\mu \partial^\mu + m]\psi - ieA_\mu \overline{\psi}\gamma^\mu\psi - \frac{1}{4}(Z_3 - 1)F^{\mu\nu}F_{\mu\nu} \\ &- (Z_2 - 1)\overline{\psi}[\gamma_\mu \partial^\mu + m]\psi + Z_2 \delta m \overline{\psi}\psi - ie(Z_2 - 1)A_\mu \overline{\psi}\gamma^\mu\psi \end{aligned} \tag{9.19}$$

where:

$$F^{\mu\nu} = \partial^\mu A_\mu - \partial^\nu A_\nu$$

with $E_i = cF_{0i}$ and $B_i = \frac{1}{2}\varepsilon_{ijk}F^{jk}$. The renormalization constants of electron mass, wavefunction and charge are δm, Z_2 and Z_3, respectively. Quantum statistical field theory provides a calculational scheme for finite temperature and density corrections using relativistic distribution functions of Bose–Einstein distribution for the possible interactions with bosons and Fermi–Dirac distributions for the possible interactions with fermions. These background contributions are incorporated as the perturbative contributions due to the interaction with the background.

These particles interact with the medium while propagating through the medium due to the virtually-emitted particles that perturb the medium. These perturbative interactions take place in the medium and contribute to modify the effective mass, charge and wavefunction of the propagating particles. Temperature and chemical potential and the high-energy radiations in the medium can produce singularities that need to be canceled out to establish the renormalization of a physically acceptable theory. For this purpose, order-by-order cancellation of singularities is required by the Kinoshita–Lee and Nauenberg (KLN) theorem and makes it possible to calculate the renormalization constant of QED. These are physically measurable parameters of the theory at each level of perturbation and contribute to electron mass, wavefunction and charge, respectively. The QED coupling is proportional to the square of electron charge, and therefore, has a measurable effect on the electromagnetic properties of the medium itself.

As discussed earlier, real-time formalism is the only way to produce the convergent perturbative series in QED. During the calculations of the renormalization conditions, we can find that QED may not work at extremely high temperatures as the incorporation of nuclear interactions is no longer unavoidable. Therefore, at high temperatures, the renormalizability of QED is not fully accepted as a standalone theory. However, the effective potential approach can incorporate some contributions which may not be pure QED contributions and show the renormalizeability. During this study, it has also been noticed that the QED is perfectly renormalizeable up to 1 billion Kelvin. Afterwards, some other contributions are needed beyond neutrino decoupling temperature which is around 2 MeV in the early universe. This remormalizeability can be re-established in QED in the presence of extremely high densities of superdense stars such as neutron stars. However, in the presence of large chemical potential, thermal effects can be counterbalanced and renormalizeabilty is re-established. However, these latest research results can only be seen in new literature and further technical details are out of the scope of this book.

At extremely high energies, the behavior of electromagnetic waves is also worth studying. The electromagnetic signal carries the information about the electromagnetic properties of the medium which may later be used to collect all the information about the stellar interiors and the early universe. The spherical nature of light is associated with the fact that electromagnetic waves satisfy the equation of a sphere created by three coordinates of electric field E, magnetic field B and the propagation vector k. For the given value of the angular velocity ω, E, B and k vectors are all perpendicular to one another in natural units such that $E^2 + B^2 + k^2 = \omega^2$. The propagation of longitudinal and transverse components in space is associated with the angular frequencies as the fourth component can change the energy because the massless energy quanta may be modified in an interacting medium and is worth studying.

9.3.4 Propagation of electromagnetic waves in QED media

The electrodynamics and quantum mechanics cannot go to the detailed study of subatomic physics for single-particle dynamics. When electromagnetic waves propagate through an interacting medium, they interact through the physical parameters of the system such as the renormalized values of electron mass, wave-function and charge in the given statistical background. The electromagnetic waves in free space exhibit only the transverse components, moving with a speed of 2.99×10^8 m s^{-1}, and have zero longitudinal components, whereas the speed of light changes due to the interaction with the medium and the refraction of light is noticed. Modifications in the properties of light in extremely hot and dense media, as observed before, should be incorporated to determine the accurate structure and composition and interpret the observational data in astronomy. The system under observation is best described by QED, which is an ultra-relativistic electron–photon system. It allows the coexistence of variable phases of fluids in astrophysical systems, locally. QED fluids are electromagnetically-interacting fluids at extremely high temperatures and densities.

9.3.5 Future application of QED in materials

QFTs are needed to describe the physics and dynamics of fundamental particles. Quantum mechanics cannot fully describe the dynamics of individual particles, especially when dealing with the light mass of electrons which can exhibit relativistic motion at very low energies. Therefore, the role of QFT in the study of high-energy physics, astrophysics and cosmology is undeniable. Actually, high-energy physics compelled scientists to develop QFT in the first place. A detailed study of nuclear physics is not possible without QFT, especially to explore the sub-nucleon level starting with the discovery of quarks. Meanwhile, heavy-ion collisions are studied using the finite temperature and density form of QCD. Understanding of radioactive decays and a detailed study of nuclear processes, even deep inelastic scattering, cannot be studied without QFT, so the development of QFT was a natural consequence of in-depth exploration of matter. Study of deep inelastic scattering, discovery of quarks and the development of QCD was impossible without QFT.

Many-body effects, quantum computing, quantum simulations, hardware development for quantum computing, quantum entanglement and the remote study of space are still challenging and QED provides a potential resource for the development of artificial intelligence to be able to study extremely large data in manageable time to understand the details. Nuclear forces may not be relevant for the study of chemical processes. However, for a detailed study of matter such as nuclear chemistry and nuclear research contribution to materials, especially to understand superconductivity, superfluidity and the existance of metamaterials, we cannot ignore nuclear forces altogether. However, it has to be remembered that gravity is not ignorable at all at macroscopic level, but it is considered ignorable at micro- and nano- scales (as of today) for all practical purposes. Once a better understanding of gravity at extremely high energies is understood, its relevance to physical problems can be investigated. It may be incorporated in the study of materials at quantum scale as well. We can expect the study of quantum materials or application of quantum field theories. Finally, we give a list of some of the well-known problems of physics and technology where QFT has to be used for better results:

1. Applications of QED in technology.
2. Energy storage.
3. Quantum optics and LASERs.
4. Quantum computing.
5. Microfluidics.
6. Biochemical physics.
7. Quantum computing and artificial intelligence.
8. System biology.
9. Heavy-ion collisions.
10. Applications to nuclear physics.
11. Astronomical study.
12. Cosmology.

However, there is still a long way to go into a detailed study of materials. The more we dig into matter, the more techniques that are needed and development of technology and mathematics will keep progressing. This is almost a continuous process and it may keep on developing as long as human interest does not shift. However, growth and development is the nature of human beings and is associated with the survival of life and may never stop.

Part IV

Appendices

IOP Publishing

Conceptual Approach to Quantum Electrodynamics

Samina S Masood

Appendix A

A.1 Fundamental units and their relations

There are basically three commonly-known systems of units and we briefly discuss them, one by one. Time is a quantity which remains the same in each and every coordinate system of units. It is usually measured in seconds, but measurements of certain quantities can use any unit such as minutes, hours, days or years. However, different quantities of time measurement are related in a standard way, even the metric units such as milliseconds and nanoseconds, whenever needed.

Mass and length are the basic fundamental units and almost all of the other quantities are derived from these basic units. The quantities used in the book will be defined mainly in well-known systems of units, including SI or MKS (m, kg, s), CGS (cm, g, s), and British Thermal Units or BTS (foot, pound, s). All the other units could be derived from three basic units of length, mass and time. The British system is distinctly different in quantum mechanics and thermodynamics. It is sometimes called the foot–pound system as the unit of distance is feet (plural of foot) and pounds are the unit of force, called pound-force. Units of the British system and metric system (SI or CGS) are converted among themselves individually. There is no standard way to relate them. The complicated mixed units involve the conversion of each individual unit correctly. For example, the mass conversion occurs as:

$$\begin{gathered} 1\ \text{kg (kilogram)} = 2.2046\ \text{lbs (pounds)} \\ 1\ \text{g} = 0.035\,274\ \text{ounces} \\ 1\ \text{lb} = 16\ \text{ounces} \end{gathered}$$

and the basic units of length can be related as:

$$\begin{gathered} 1\ \text{m (meter)} = 3.280\,84\ \text{ft (feet)} \\ 1\ \text{km} = 0.621\,37\ \text{mile} \\ 1\ \text{mile} = 5280\ \text{ft} \end{gathered}$$

The unit of energy is called BTU and temperature is defined in Fahrenheit that is set to 32 for the freezing point of water and 212 for the boiling point, and the relation between Celsius or centigrade with Fahrenheit is defined as:

doi:10.1088/978-0-7503-6054-8ch10

$$100°C = 180°F$$
$$0°C = 32°F; \ 100°C = 212°F$$
$$°C = 100(°F - 32)/180$$

whereas, the units for power and energy are related in different ways as follows:

$$1 \text{ BTU (British Thermal Unit)} = 1.055 \times 10^3 \text{ Joules}$$

Another relatively common distinct unit of the British system is called horsepower. It is the power needed to pick up 550 pounds to a height of 1 foot in one second. The British system defines every individual unit for the same quantity in a different way so there was a lot to remember when more parameters were defined. Several countries including the United States use the British system in their daily life. However, the scientific community is mainly using the metric system as common units. MKS (meter–kilogram–second) is used for larger systems and CGS (centimeter–gram–second) system is used for tiny systems, but the conversion among different units for the same quantity can be done by multiplying factors in powers of ten, which are tabulated in table A.1.

The metric system is much simpler, as compared to the British system. The connection between the small and large quantities is defined individually in this system and each quantity has a different name. Just to give an example: inch, foot, yard, furlong and mile are all units of length and are connected through individual conversion factors. We will not discuss these units in further detail. The international common units are metric units. They are convenient and are related together through a standard conversion system. Just a prefix is used to relate all quantities through the metric system for the same quantity. All of these prefixes are listed in table A.1 and are expressed in powers of ten, hence named as a metric system.

The commonly called SI (Systeme Internationale) system is the international system of measurement. It uses the meter as the unit for length, whereas the mass is measured in kilograms, and time in seconds. SI units are the convenient units and

Table A.1. Prefixes in metric systems.

Prefix	Metric small unit	Prefix	Metric big unit
Yocto	10^{-24}	Yotta	10^{24}
Zepto	10^{-21}	Zeta	10^{21}
Atto	10^{-18}	Exa	10^{18}
Femto	10^{-15}	Peta	10^{15}
Peco	10^{-12}	Tera	10^{12}
Nano	10^{-9}	Giga	10^{9}
Micro	10^{-6}	Mega	10^{6}
Mili	10^{-3}	Kilo	10^{3}
Centi	10^{-2}	Hecta	10^{2}
Deci	10^{-1}	Deca	10^{1}
Actual Unit	1	Actual Unit	1

are adopted in several countries as a daily-life measurement system instead of the British system. It is also called the meter–kilogram–second (MKS) system. Another form of metric system is called Gaussian units or the centimeter–gram–second (CGS) system. The CGS system is more convenient to handle small quantities and conversion in between two metric systems is conveniently done using the powers of ten. The metric system develops a direct relationship among small and big quantities through powers of ten and general prefixes are used for that purpose. Table A.1 shows these prefixes and they can be attached to any quantity to express any small and big quantities in terms of one standard unit.

The Gaussian system or CGS system of units gives another set of units for small scale quantities in metric system and can be directly converted to SI units at any stage by multiplying with powers of ten at any point. SI units are a straightforward extension of Gaussian units. Electromagnetism mainly uses the metric system at the international level. Coulomb's force and the electric and magnetic fields are related to the intrinsic behavior of charge. Electric and magnetic fields are dependent on the displacement and the velocity of charge. Dynamics of charges cannot be described without current, the flow of charge, which may be defined as the product of charge and the velocity of charge. So, the rate of flow of charge or current describes the dynamics instead of velocity as it can incorporate the variation of associated electric and magnetic fields with charge as well. Basic principles of classical physics are based on mass, velocity, acceleration and force, whereas additional parameters of the theory appear in electromagnetism without getting rid of the basic parameters of mechanics and make the theory much more complicated.

The main parameters of electromagnetism include charge, electric and magnetic field, resistance, current and voltage. A few more complex parameters such as capacitance, conductance, voltage are for the measure of potential energy per unit charge and their unit is called volts or joules per coulomb. There are several other parameters of electromagnetic theory that are expressed in terms of the basic units of electromagnetism. Some of the well-known parameters are related to properties of materials through which the flow of charge takes place in the form of current. Electromagnetic properties of materials such as conductivity, resistivity, electric permittivity, magnetic permeability, dielectric constant and refractive index are defined in terms of the known quantities.

For example, the Coulomb constant $k = \frac{1}{4\pi\varepsilon_0}$ is given in SI units, where ε_0 is the electric permittivity in free space and is given as $\varepsilon_0 = 8.854\ 287\ 82\ \mathrm{C\ (Nm)^{-2}}$. The speed of light in free space is related to electric permittivity as $c = \frac{1}{\sqrt{(\varepsilon_0 \mu_0)}}$ such that $\mu_0 = 4\pi \times 10^{-7}$ newton per square ampere or $\mu_0 = 1.26 \times 10^{-6}$ weber per ampere-meter (newton/square ampere). Weber is the unit of electric flux in SI units defined as tesla per meter square. Magnetic flux is the number of lines of magnetic field per unit area. Table A.2 gives a few well-known quantities in various systems of units.

All of the above systems of units work well, in general. For scientific purposes, the metric system is considered much more convenient and is accepted internationally to share scientific discoveries and technological development.

Table A.3 can be used to convert each of the individual units.

Table A.2. Quantities in different system of units ESU (electrostatic unit).

Quantities	SI (MKS)	Gaussian (CGS)	British (FPS)
Distance	meter (m)	centimeter (cm)	foot (ft)
Mass	kilogram (kg)	gram (g)	slug
Time	second (s)	second (s)	second (s)
Velocity	m s^{-1}	cm s^{-1}	ft s^{-1}
Acceleration	m s^{-2}	cm s^{-2}	ft s^{-2}
Force	newton (N)	dynes (Dy)	pound-force
Energy	joule (J)	ergs (ergs)	pound-energy
Heat	joule	calorie	BTU
Temperature	Celsius	centigrade	Fahrenheit
Absolute temperature	kelvin	Kelvin	Rankine
Charge	coulomb	statcoulomb	*ESU
Voltage	volt	statvolt	...
Electric current	ampere	statampere	...
Conductance	mho	statmho	...
Resistance	ohms	statohms	...
Magnetic flux	weber	maxwell	...
Amount of substance	gram-mole	gram-mole	pound-mole

Table A.3. Values of a few parameters in SI units.

Quantities	Symbol	SI (MKS)
Charge of electron	e	1.602×10^{-19}
Speed of light in space	c	2.998×10^{8} m s^{-1}
Planck constant	h	6.626×10^{-34} J s
Boltzmann constant	k_B	1.381×10^{-23} J K^{-1}
Coulomb constant	k	8.987×10^{9} kg m^{3} s^{-2} C^{-2}
Electric permittivity	ε_0	8.854 C^2N m^2
Magnetic Permeability	μ_0	$4\pi \times 10^{-7} = 1.256 \times 10^{-6}$ Tm A^{-1}

A.1.1 Natural system of units

In high energy physics, the relativistic velocities and high energies lead to another more convenient system where we can reduce the commonly used constants by setting them equal to 1 to define a new set of working units that are directly related to the extremely high energy. In this system of units, the commonly appearing constants are absorbed in the newly defined units for practical purposes. When conversion into well-known coordinates is needed, the values of coefficients can be included back. So, we keep all the values of coefficients set to unity in the standard units as $\hbar = c = k_B = 1$, where h (or $\hbar$) is Planck's constant, c is speed of light and k_B is the Boltzmann constant.

In the natural system of units, distances are measured in femtometers (fm) and energy in million electronvolt (MeV). Distance and energy are considered to be two fundamental units to describe most of the quantities. Then, relativity allows one to use the relationship between energy, momentum and mass. The conversion units can be used in the end to change energy into joules for SI or any other known energy unit. Particle physics uses relativity and quantum mechanics, in the rest frame of particles, simultaneously. Therefore, wave–particle duality allows treating particles' motion as wave propagation along with the energy–momentum relations of relativity. A few fundamental relativistic relations are used to convert the coordinates properly:

$$E^2 = p^2c^2 + m^2c^4 \tag{A.1}$$

$$E = mc^2 = pc \tag{A.2}$$

$$E = \hbar\omega \tag{A.3}$$

$$p = mc = \hbar k = \frac{h}{\lambda} \tag{A.4}$$

$$k = \frac{2\pi}{\lambda} \tag{A.5}$$

in usual notation of wave mechanics, and taking $\hbar = c = k_{\mathrm{B}} = 1$, energy, momentum, and mass can all be expressed in units of MeV. Even the length comes out to be an inverse of mass and is sometimes expressed as the reciprocal of energy 1/MeV. This length is related to the femtometer scale as well. Standard values of a few well-known constants are given in table A.4.

Table A.4. Values of a few important constants.

Quantity	Values (SI units)	Values (natural units)
$\hbar$	1.055×10^{-34} J S	1
k_{B}	1.38×10^{-23} J K^{-1}	1
c	2.99×10^{8} m s^{-1}	1
Energy	Joules	MeV
MeV	1.6×10^{-13} J	1
Mass	kg	MeV/c^2
Momentum	kg m s^{-1}	MeV/c
Electron mass	9.109×10^{-31} kg	0.511 MeV/c^2
Proton mass	1.673×10^{-27} kg	938.272 MeV/c^2
Neutron mass	1.675×10^{-27} kg	939.565 MeV/c^2
α	$\frac{e^2}{4\pi\varepsilon_0 \hbar c}$	$\frac{e^2}{4\pi}$

A.2 Laws of electromagnetism

There are four basic laws of electrodynamics which make up Maxwell's equations that allow one to express electromagnetic waves in terms of electric and magnetic fields. Four basic principles of electrodynamics are briefly discussed here to develop Maxwell's equations. These laws include Gauss's law of electric and magnetic fields and then Faraday's law and Ampère's law. All of the other equations of electrodynamics can be derived from these four basic principles of electromagnetism and are found through experiments.

A.2.1 Gauss's law

Electric charge is associated with the electric field and the amount of charge per unit volume is defined as electric charge density. Since we do not observe magnetic monopoles, the magnetic charge density is always zero. If ρ is defined as free charge density then the variation in the electric field vector is given in terms of **Gauss's law** and is written as:

$$\nabla \cdot \mathbf{E} = 4\pi\rho \tag{A.6}$$

the corresponding expression for the magnetic field is expressed as:

$$\nabla \cdot \mathbf{B} = 0 \tag{A.7}$$

due to the vanishing of magnetic free charge density as magnets are always found as magnetic dipoles, and magnetic monopoles do not exist. This indicates the major difference between electricity and magnetism. Charge is the intrinsic property of matter, whereas magnetic poles are associated with the motion of charge and the phenomenon of magnetism is associated with the dynamics of charges.

A.2.2 Faraday's law of magnetic induction

The variation in the magnetic field induces a voltage in the circuit. If this variation takes place in a coil, the voltage induced in the coil of wire is called induced electromotive force (emf) that increases or decreases the voltage in the coil. The amount of induced voltage depends on the frequency of variation of the field. This change in voltage depends on the strength of the magnetic field as well as the change in the field.

$$\nabla \times \mathbf{E} = -\frac{1}{c}\frac{\partial \mathbf{B}}{\partial t} \tag{A.8}$$

If a constant electric current passes through a coil, the strength of magnetic field will depend on the number of turns in the coil. The magnetic flux in the loop constant is the induced magnetic field inside any loop of a wire. **Faraday's law:** Any change in the magnetic environment of a coil of wire will cause a voltage (emf) to be 'induced' in the coil. No matter how the change is produced, the voltage will be generated. The change could be produced by changing the magnetic field strength, moving a magnet towards or away from the coil, moving the coil into or out of the magnetic field, rotating the coil relative to the magnet, etc. Faraday's law is a fundamental relationship which comes from Maxwell's equations. It serves as a succinct summary of the ways a voltage (or emf) may be generated by a changing the magnetic environment. The induced emf in a coil is equal to the negative of the rate of change of magnetic flux times the number of turns in the coil. It involves the interaction of charge with magnetic field. When an emf is generated by a change in magnetic flux according to Faraday's Law, the polarity of the induced emf is such that it produces a current whose magnetic field opposes the change which produces it. The induced magnetic field inside any loop of wire always acts to keep the magnetic flux in the loop constant. In the examples below, if the field B is increasing, the induced field acts in opposition to it. If it is decreasing, the induced field acts in the direction of the applied field to try to keep it constant.

A.2.3 Ampère's circuital law

The integral form of Ampère's circuital law for magnetostatics relates the magnetic field perpendicular in a circular wire with the rate of change of electric field and along a closed path to the total current flowing through any surface bounded by that path. This law can be considered as a complimentary law to Faraday's law where the variation in electric field and the current both contribute to produce some magnetic fields.

$$\nabla \times \mathbf{B} = \frac{4\pi}{c}\mathbf{J} + \frac{1}{c}\frac{\partial \mathbf{E}}{\partial t} \tag{A.9}$$

A.3 Maxwell's equations

The most important fundamental laws of electrodynamics are combined together as Maxwell's equations. These equations were used to develop the electromagnetic wave theory and showed that the laws of electromagnetism satisfy a wave equation. A detailed study of electromagnetic signals was prompted at this point which led to the four-dimensional representation of Maxwell's equations and finally helped in the development of QED.

We first describe Maxwell's equations in free space where ε_0 and μ_0 give the electric permittivity and magnetic permeability of free space with its constant values. This also gives the constant value to the speed of light as $c = (\sqrt{1/(\varepsilon_0 \mu_0)})$. The differential form of Maxwell's equations in free space (in the standard SI units) is written as:

$$\nabla \cdot \mathbf{B} = 0 \tag{A.10}$$

$$\nabla \cdot \mathbf{E} = \frac{\rho}{\varepsilon_0} \tag{A.11}$$

$$\nabla\times\mathbf{E} = -\frac{\partial \mathbf{B}}{\partial t} \tag{A.12}$$

$$\nabla\times\mathbf{B} = \mu_0\left(\mathbf{J} + \varepsilon_0\frac{\partial \mathbf{E}}{\partial t}\right) \tag{A.13}$$

The corresponding integral form of the Maxwell's equations in free space (SI units) is written as:

$$\oint \mathbf{E}\cdot\mathbf{dS} = \frac{1}{\varepsilon_0}\iiint_\Omega \rho dV \tag{A.14}$$

$$\oint \mathbf{B}\cdot\mathbf{dS} = 0 \tag{A.15}$$

$$\oint \mathbf{E}\cdot\mathbf{dl} = -\frac{d}{dt}\iint \mathbf{B}\cdot\mathbf{dS} \tag{A.16}$$

$$\oint \mathbf{B}\cdot\mathbf{dl} = \mu_0\left(\iint \mathbf{J}\cdot\mathbf{dS} + \varepsilon_0\frac{d}{dt}\iint_\Sigma \mathbf{E}\cdot\mathbf{dS}\right) \tag{A.17}$$

The two systems of units in metric systems are very similar and easily translated in metric notation. The SI system is more generalized and its scope is big. It works from the largest to the smallest scale at the same time. It covers from micro or nano scales to cosmic scales. However, the values of constants in classical electrodynamics are very convenient in the Gaussian or CGS system because we can set the Coulomb constant $k = 1/4\pi\varepsilon_0 = 1$ and then the integral form of Maxwell's equations in Gaussian units attain the simpler form as:

$$\oint \mathbf{E}\cdot\mathbf{dS} = 4\pi\iiint_\Omega \rho dV \tag{A.18}$$

$$\oint \mathbf{B}\cdot\mathbf{dS} = 0 \tag{A.19}$$

$$\oint \mathbf{E}\cdot\mathbf{dl} \cdot -\frac{1}{c}\frac{d}{dt}\iint \mathbf{B}.\mathbf{dS} \tag{A.20}$$

$$\oint \mathbf{B}\cdot\mathbf{dl} = \frac{4\pi}{c}\iint \mathbf{J}\cdot dS + \frac{1}{c}\frac{d}{dt}\iint \mathbf{E}\cdot\mathbf{dS} \tag{A.21}$$

and the corresponding differential form can be written as:

$$\nabla\cdot\mathbf{B} = 0 \tag{A.22}$$

$$\nabla\cdot\mathbf{E} = 4\pi\rho \tag{A.23}$$

$$\nabla\times\mathbf{E} = -\frac{1}{c}\frac{\partial \mathbf{B}}{\partial t} \tag{A.24}$$

$$\nabla \times \mathbf{B} = \frac{4\pi}{c}\mathbf{J} + \frac{1}{c}\frac{\partial E}{\partial t} \tag{A.25}$$

These units are very convenient when we translate Maxwell's equations in a medium and define the displacement vector D, which corresponds to $D = \varepsilon_0 \varepsilon_r E$ in terms of the relative permittivity (ε_r) and permittivity in free space (ε_0) such that they together indicate the total permittivity ε of a medium. Similarly, the magnetic field in the medium is defined as H and $H = B/\mu_0$. This helps to conveniently define magnetization, polarization and other vectors in a more convenient way and SI units can be retrieved easily, whenever needed. Some of the important units are tabulated at the end for comparison, and make the transformation of units easy for calculation.

Electric and magnetic fields are usually referred to as classical fields because they are three-dimensional fields and can attain any value. So, these fields are considered as continuous variables because they are associated with charges which are composite charges (quantized) which produce the continuous values of fields due to the continuous variation of the separation of charges.

A.4 Properties of γ-matrices

The Pauli matrices are a special case in the Clifford algebra, which is the algebra generated by the product of vector spaces. Going from quadratic to linear vector space, an extra factor is used to identify the quadratic form distinguished by unital associative algebra. It represents the physical space, $C(3, 0)$, which is called Pauli algebra. The two-dimensional representation of a Lorentz group corresponds to spin $s = 1/2$. These matrices are the 2×2 unitary matrices with a unit determinant I and can be expressed as:

$$U = e^{i\theta^i \sigma^i/2}$$

where θ^i are three arbitrary parameters and σ^i are the Pauli spin matrices. The γ-matrices are the four-dimensional generalization of the two-dimensional Pauli matrices σ^i discussed and they are significant in the formulation of the Dirac equation.

There are two important requirements for the Dirac equation. The first is that it has to be a first-order equation in space and time coordinates. The second requirement is that it has to eliminate the negative probability density arising from the Klein–Gordon equation. So the Klein–Gordon equation was modified as:

$$(\partial_\mu \partial^\mu + m^2)\psi = 0$$

Applying $i\gamma^\nu \partial_\nu$ on both sides to the first-order equation gives

$$-\gamma^\nu \gamma^\mu \partial_\nu \partial_\mu \psi = im\gamma^\nu \partial_\nu \psi = m^2 \psi$$

Due to the tensor properties, this can be written as

$$-\frac{1}{2}(\gamma^\nu \gamma^\mu + \gamma^\mu \gamma^\nu)\partial_\nu \partial_\mu \psi = m^2 \psi$$

The term in the parenthesis is the anticommutator $\{\gamma^\nu, \gamma^\mu\} = \gamma^\nu\gamma^\mu + \gamma^\mu\gamma^\nu$. If we consider a plain wave $\psi e^{ip^\mu x_\mu}$ and since $p_\mu p^\mu = m^2$, the Dirac equation is

$$\frac{1}{2}\{\gamma^\nu, \gamma^\mu\}p_\nu p_\mu = m^2.$$

We can clearly see that the anticommutator $\{\gamma^\nu, \gamma^\mu\} = 2\eta^{\nu\mu}$, where $\eta^{\nu\mu}$ are the components of a 4×4 matrix, Minkowski metric, and therefore the four quantities γ^μ also have to be 4×4 matrices. From the obtained relation, we can observe some properties of the gamma matrices, also known as Dirac matrices.

$$(\gamma^0)^2 = \eta^{00} = 1,\ (\gamma^i)^2 = -1 \tag{A.26}$$

$$\gamma^0\gamma^i + \gamma^i\gamma^0 = 0, \tag{A.27}$$

$$\gamma^i\gamma^j + \gamma^j\gamma^i = 2\delta^{ij} \tag{A.28}$$

and $\gamma^j \equiv i\sigma^j$. For $\nu \neq \mu$:

$$\gamma^\mu\gamma^\nu = -\gamma^\mu\gamma^\nu$$

Now set $\nu = 0$ and $\mu = i$:

$$\gamma^0\gamma^i = 2g^{\mu\nu} - \gamma^i\gamma^0$$

Multiplying by γ^i on both sides and using the identity $\gamma^i\gamma^i = \delta^{ii} = 1$:

$$\gamma^0 = \gamma^i\gamma^0\gamma^i \tag{A.29}$$

Using cyclic rotation of tensors, the trace of the results can be observed as:

$$Tr(\gamma^0) = Tr(\gamma^i\gamma^0\gamma^i) = Tr(\gamma^0) = 0 \tag{A.30}$$

The matrix representations are:

$$\gamma^0 = \begin{pmatrix} \mathbb{I} & 0 \\ 0 & \mathbb{I} \end{pmatrix} \tag{A.31}$$

where $\mathbb{I}$ is the 2×2 unit matrix.

$$\gamma^i = \begin{pmatrix} 0 & \sigma^i \\ -\sigma^i & 0 \end{pmatrix} \tag{A.32}$$

This representation is known as Weyl or Chiral representation.

A.5 Feynman rules in QED

QED lagrangian is given as

$$\mathscr{L}_{\mathscr{QED}} = -\frac{1}{4}F_{\mu\nu}F^{\mu\nu} + \bar{\psi}(i\gamma_\mu\partial_\mu - m)\psi + \bar{\psi}(-ie\gamma^\mu)\psi A_\mu \tag{A.33}$$

A.6 Feynman propagators

Feynman propagators are defined differently for charges of fermions and photons:

- For each vertex, write a factor γ^{α}.
- For each internal photon line, labeled by the momentum k, write a factor

$$iD_{F_{\alpha\beta}}(k) = i\frac{-g_{\alpha\beta}}{k^2 + i\varepsilon}$$

  A boson line is represented as a wavy line and arrow distinguishing between the incoming and the outgoing boson. Even the propagating boson is also represented by a straight wavy line. A summary of Feynman rules is shown in figure A.1. The fermion line is represented as a straight line as virtual fermion or an incoming or outgoing fermion.
- For each internal fermion line, labeled by the momentum p, write a factor

$$iS_F(p) = i\frac{1}{\not{p} - m + i\varepsilon}$$

  where **p** and **k** denote the three-momenta of the external particles and r (=1, 2) label their spin and polarization states.
- The spinor factors (γ-matrices, S_F-functions, four-spinors) for each fermion line are ordered so that, reading from right to left, they occur in the same sequence as following the fermion line in the direction of its arrows.
- For each closed fermion loop, take the trace and multiply by a factor (-1).
- The four-momenta associated with the three lines meeting at each vertex satisfy energy-momenta conservation. For each four-momenta q which is not fixed by energy–momentum conservation, carry out the integration

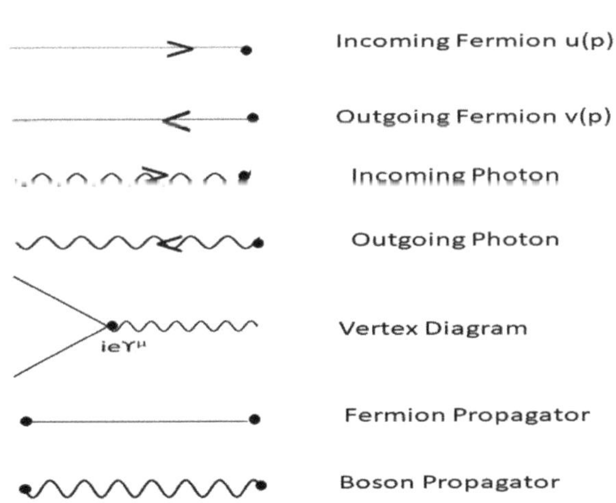

Figure A.1. Feynman rules of QED. Black dot indicates the target.

$(2\pi)^{-4} \int d^4q$. One such integration with respect to an internal momentum variable q occurs for each closed loop.

- Multiply the expression by a phase factor δ_p, which is equal to $+1(-1)$ if an even (odd) number of interchanges of neighboring fermion operators is required to write the fermion operators in the correct normal order.

Further reading

NOTE: Most of the material in this book can be partially found in various standard textbooks in more details and is summarized here. Students can use their own graduate textbooks of the relevant courses study. However, for detailed develop technical skill. Designated books are not available for the last chapter. Therefore, a few journal articles and relatively recent publications are suggested to start literature survey and familiarize with the new approach to study the application of quantum electrodynamics and quantum statistical electrodynamics.

It is also worth-mentioning that a couple of physics books which comes in series are very helpful for conceptual development (e.g., Feynman lectures) in various topics of physics and practicing problem solving (e.g., Schaum's Outline series). Berkley Series and Landau and Lifshitz's sets of books are also worth-mentioning to develop or review various topics of physics or improve the understanding.

References

1 Vectors and tensors

[1] Arfken G B and Weber H J 2005 *Mathematical Methods for Physicists.* 6th edn (Amsterdam: Elsevier)
[2] Sharipov R 2004 Quick introduction to tensor analysis (arXiv:math/0403252)
[3] Kay D 2011 *Tensor Calculus* (Schaum's Outline Series) 2nd edn (New York: McGraw-Hill)
[4] Spiegal M and Lipschtuz S 2009 *Vector Calculus* (Schaum's Outline Series) 2nd edn (New York: McGraw-Hill)

2 Differential equations and Lagrangian formalism

[5] Nakahara M 2018 *Geometry, Topology and Physics* (Boca Raton, FL: CRC Press)
[6] Fortney J P *et al* 2018 *A Visual Introduction to Differential Forms and Calculus on Manifolds* vol 2 (Berlin: Springer)

doi:10.1088/978-0-7503-6054-8ch11

[7] Boas M L 2006 *Mathematical Methods in the Physical Sciences* 3rd edn (New York: Wiley)
[8] Arnold V I and Silverman J H 1994 *Advanced Topics in the Arithmetic of Elliptic Curves* vol 151 (New York: Springer)
[9] Tenenbaum M and Pollard H 1985 *Ordinary Differential Equations: An Elementary Textbook for Students of Mathematics, Engineering, and the Sciences* (North Chelmsford, MA: Courier Corporation)
[10] Henneaux M 1982 Equations of motion, commutation relations and ambiguities in the Lagrangian formalism *Ann. Phys., NY* **140** 45–64
[11] Duren Jr W L 1965 *Basic Library List* (ERIC)
[12] Clark C W 1963 *Ordinary Differential Equations, by Morris Tenenbaum and Harry Pollard* (New York: Harper and Row)
[13] Wrede R and Spiegal M S 2010 *Advanced Calculus* 3rd edn (New York: McGraw-Hill)

3 Computational tools

[14] Luchtenburg D M 2021 Data-driven science and engineering: machine learning, dynamical systems, and control *IEEE Control Syst. Mag.* **41** 95–102
[15] Kong Q, Siauw T and Bayen A 2020 *Python Programming and Numerical Methods: A Guide for Engineers and Scientists* (New York: Academic)
[16] Cormen T H, Leiserson C E, Rivest R L and Stein C 2009 *Introduction to Algorithms* 3rd edn (Cambridge, MA: MIT Press)
[17] Teukolsky S A, Vetterling W T and Flannery B P 2007 *Numerical Recipes: The Art of Scientific Computing* ed Press W H (Cambridge: Cambridge University Press)
[18] Stoer J and Bulirsch R 2002 *Introduction to Numerical Analysis* 3rd edn (Berlin: Springer)
[19] Lipschutz S and Lipson M L 1986 *SCHAUM'S: Outline of Theory and Problems of Data Structures* (New York: McGraw-Hill)

4 Electromagnetism

[20] Milton K and Schwinger J 2024 *Classical Electrodynamics* (Boca Raton, FL: CRC Press)
[21] Griffiths D J 2023 *Introduction to Electrodynamics* (Cambridge: Cambridge University Press)
[22] Jackson J D 2021 *Classical Electrodynamics* (New York: Wiley)
[23] Walter G 2012 *Classical Electrodynamics* (Berlin: Springer Science +Business Media)
[24] Feynman R P, Leighton R B, and Sands M 1964–5 *The Feynman Lectures on Physics* vols 1–3 (Reading, MA: Addison-Wesley)

5 Thermodynamics and statistical mechanics

[25] Shell M S 2015 *Thermodynamics and Statistical Mechanics: An Integrated Approach* (Cambridge: Cambridge University Press)

[26] Pathria R K and Beale P D 2011 *Statistical Mechanics*. 3rd edn (Amsterdam: Elsevier)
[27] Roy B N 2002 *Fundamentals of Classical and Statistical Thermodynamics* (New York: Wiley)
[28] Huang K 1987 *Statistical Mechanics*. 2nd edn (New York: Wiley)
[29] Reif F 1965 *Fundamentals of Statistical and Thermal Physics* (New York: McGraw-Hill)

6 Quantum mechanics

[30] Griffiths D J and Schroeter D F 2018 *Introduction to Quantum Mechanics* (Cambridge: Cambridge University Press)
[31] Sakurai J J and Napolitano J 2017 *Modern Quantum Mechanics* (London: Pearson)
[32] Shankar R 2012 *Principles of Quantum Mechanics* (Berlin: Springer Science+Business Media)
[33] Zettili N 2009 *Quantum Mechanics: Concepts and Applications* 2nd edn (New York: Wiley)
[34] Gautreau R and Savin W 1997 *Schaum's Outline of Modern Physics* 2nd edn (New York: McGraw-Hill)
[35] Purcell E M *et al Berkeley Physics Course* vol 1: Mechanics (1960), vol 2: Electricity and Magnetism (1962), etc
[36] Landau L D and Lifshitz E M 1958 *Course of Theoretical Physics, Vol 1: Mechanics*; series continues 1959–71

7 High energy physics and relativity

[37] Schutz B 2022 *A First Course in General Relativity* (Cambridge: Cambridge University Press)
[38] Griffiths D 2020 *Introduction to Elementary Particles*, (New York: Wiley)
[39] Thomson M 2013 *Modern Particle Physics* (Cambridge: Cambridge University Press)
[40] Perkins D H 2000 *Introduction to High Energy Physics* (Cambridge: Cambridge University Press)
[41] Kalman G 1961 Lagrangian formalism in relativistic dynamics *Phys. Rev.* **123** 384

8 Quantum electrodynamics

[42] Le Bellac M 2011 *Thermal Field Theory* (Cambridge: Cambridge University Press)
[43] Peskin M E 2018 *An Introduction to Quantum Field Theory* (Boca Raton, FL: CRC Press)
[44] Schwartz M D 2014 *Quantum Field Theory and the Standard Model* (Cambridge: Cambridge University Press)
[45] Mandl F and Shaw G 2010 *Quantum Field Theory* 2nd edn (New York: Wiley)

[46] Randall L and Schwartz M D 2001 Quantum field theory and unification in AdS5 *J. High Energy Phys.* **2001** 003

[47] Fujikawa K and Shrock R E 1980 Anomalous magnetic moment of a massive neutrino and neutrino-spin rotation *Phys. Rev. Lett.* **45** 963

[48] Bjorken J D and Drell S D 1965 *Relativistic Quantum Fields* (New York: McGraw-Hill)

[49] Bjorken J D and Drell S D 1964 *Relativistic Quantum Mechanics* (New York: McGraw-Hill)

[50] Kinoshita T 1962 Mass singularities of Feynman amplitudes *J. Math. Phys.* **3** 650

[51] Kalman G 1961 Lagrangian formalism in relativistic dynamics *Phys. Rev.* **123** 384

[52] Dirac P A M, Fock V A and Podolsky B 1943 *Quantum Electrodynamics. No. 1* (Dublin: Dublin Institute for Advanced Studies)

9 Applications of quantum electrodynamics

[53] Caliskan S and Masood S 2025 A first principles study on spin dependent electronic and optical characteristics of NH_2 adsorbed zinc oxide-based nanomaterials *Physica B* **706** 417142

[54] Masood S and Singh J 2025 Thermal contributions to primordial nucleosynthesis *Int. J. Mod. Phys. A* **25** 50043

[55] Caliskan S, Mammadov A and Masood S 2023 Spin dependent electronic characteristics of zinc oxide nanowires linked to nickel electrodes: a first principles study *Solid State Commun.* **369** 115211

[56] Fujii Y *et al* 2022 Applications of QED corrections in collider physics, *Eur. Phys. J. C*; also resources from CERN and DESY websites

[57] Masood S and Caliskan S 2022 Equation of state of fermions in neutron stars (arXiv:2206.09486)

[58] Masood S and Mein H 2022 Comparative study of magnetic moment of leptons in hot and dense media *Int. J. Mod. Phys. A* **37** 2250120

[59] Masood S S 2020 *QED at Finite Temperature and Density* (Singapore: World Scientific)

[60] Masood S S 2019 QED plasma in the early universe *Arab. J. Math.* **8** 183–92

[61] Saleem I *et al* 2018 Adhesion of gram-negative rod-shaped bacteria on 1D nano-ripple glass pattern in weak magnetic fields *MicrobiologyOpen* **8** e640

[62] Masood S 2017 Preliminary study of the effects of magnetic field on bacteria *Biophys. Rev. Lett.* **177** 1–12

[63] Masood S S 2017 Propagation of monochromatic light in a hot and dense medium *Eur. Phys. J. C* **77** 826

[64] Masood S and Saleem I 2017 Propagation of electromagnetic waves in extremely dense media *Int. J. Mod. Phys. A* **32** 1750081

[65] Masood S S and Miller A 2014 A von Neumann entropy measure of entanglement transfer in a double Jaynes–Cummings model (arXiv:1412.5410)
[66] Masood S and Miller A 2014 Entanglement in a Jaynes–Cummings model with two atoms and two photon modes *Universal J. Phys. Appl.* **2** 237–44
[67] Masood S and Haseeb M 2012 Second order corrections to anomalous magnetic moment of electron at finite temperature *Int. J. Mod. Phys. A* **27** 1250188
[68] Haseeb M and Masood S S 2011 Second order thermal corrections to electron wavefunction *Phys. Lett. B* **704** 66
[69] Masood S and Haseeb M 2011 Two loop temperature corrections to electron self-energy *Chin. Phys. C* **35** 608
[70] Nielsen M A and Chuang I L 2011 *Quantum Computation and Quantum Information.* 10th anniversary edn (Cambridge: Cambridge University Press)
[71] Masood S 2010 Quantum electrodynamics of nanosystems *NSTI-Nanotech* **1** 249
[72] Eidelman S and Passera M 2007 Theory of the tau lepton anomalous magnetic moment *Mod. Phys. Lett. A* **22** 159–79
[73] Kapusta J I and Gale C 2006 *Finite-Temperature Field Theory: Principles and Applications* 2nd edn (Cambridge: Cambridge University Press)
[74] Masood S, Perez Martinez A, Perez Rojas H, Gaitan R and Rodriguez Romo S 2002 Effective anomalous magnetic moment of neutrinos in the strong magnetic fields *Rev. Mex. Fis.* **48** 501
[75] Masood S *et al* 2002 Effective magnetic moment of neutrinos in the strong magnetic fields *Rev. Mex. Fis.* **48** 501
[76] Czarnecki A and Marciano W J 2001 Muon anomalous magnetic moment: a harbinger for 'new physics' *Phys. Rev. D* **64** 013014
[77] Chaichian M, Masood S, Montonen C, Martinez A P and Rojas H P 2000 Quantum magnetic collapse *Phys. Rev. Lett.* **84** 5261
[78] Esposito S, Mangano G, Miele G and Pisanti O 1998 Wave function renormalization at finite temperature *Phys. Rev. D* **58** 105023
[79] Chapman I A 1997 Finite temperature wave-function renormalization: a comparative analysis *Phys. Rev. D* **55** 6287–91
[80] Elmfors P, Grasso D and Raffelt G 1996 Neutrino dispersion in magnetized media and spin oscillations in the early universe (arXiv:hep-ph/9605250)
[81] Raffelt G G 1996 *Stars as Laboratories for Fundamental Physics* (Chicago, IL: University of Chicago Press)
[82] Sawyer R F 1996 Temperature-dependent wave function renormalization and weak interaction rates prior to nucleosynthesis *Phys. Rev. D* **53** 4232–6
[83] Masood S 1995 Anomalous magnetic moment of neutrino in the statistical background *Astropart. Phys.* **4** 189

[84] Masood S 1995 Magnetic moment of neutrino in the statistical background *Astropart. Phys.* **4** 189

[85] Masood S 1993 Neutrino physics in hot and dense media *Phys. Rev. D* **48** 3250

[86] Masood S 1993 Renormalization of QED in superdense media *Phys. Rev. D* **47** 648

[87] Aurenche P and Becherrawy T 1992 A comparison of the real-time and the imaginary-time formalisms of finite-temperature field theory for 2, 3 and 4-point Green functions *Nucl. Phys. B* **379** 259–303

[88] Masood S and Haseeb M 1992 Second order electron mass dispersion relation at finite temperature-II *Phys. Rev. D* **46** 5633

[89] Ahmed K and Masood S S 1991 Vacuum polarization at finite temperature and density in QED *Ann. Phys.* **164** 460

[90] Masood S S 1991 Photon mass in the classical limit of finite-temperature and -density QED *Phys. Rev. D* **44** 3943

[91] Masood S and Haseeb M 1991 Second order electron mass dispersion relation at finite temperature *Phys. Rev. D* **44** 3322

[92] Pal P B and Mohapatra R N 1991 *Massive Neutrinos in Physics and Astrophysics* (Singapore: World Scientific)

[93] Kobes R 1990 A correspondence between imaginary time and real time finite temperature field theory *Phys. Rev. D* **42** 562

[94] Nieves J F and Pal P B 1989 P- and CP-odd terms in the photon self-energy within a medium *Phys. Rev. D* **39** 652

[95] Ahmed K and Masood S S 1987 Finite-temperature and -density renormalization effects in QED *Phys. Rev. D* **35** 4020

[96] Ahmed K and Masood S S 1987 Renormalization and radiative corrections at finite temperature reexamined *Phys. Rev. D* **35** 1861

[97] Babu K S and Mathur V S 1987 Radiative corrections to neutrino masses *Phys. Lett. B* **196** 218

[98] Fukugita M and Yanagida T 1987 Particle-physics model for Voloshin-Vysotsky-Okun solution to the solar-neutrino problem *Phys. Rev. Lett.* **58** 1807

[99] Landsman N P and van Weert C G 1987 Real- and imaginary-time field theory at finite temperature and density *Phys. Rep.* **145** 141 and references therein

[100] Masood S S 1987 Finite-temperature and -density effects on electron self-mass and primordial nucleosynthesis *Phys. Rev. D* **36** 2602

[101] Kobes R L and Semenoff G W 1986 Discontinuity of Green functions in field theory at finite temperature and density (II) *Nucl. Phys. B* **272** 329–64

[102] Donoghue J F, Holstein B R and Robinett R W 1985 Quantum electrodynamics at finite temperature *Ann. Phys.* **164** 233–7

[103] Levinson E J and Boal D H 1985 Relativistic finite temperature field theory *Phys. Rev. D* **31** 3280

[104] Donoghue J F and Holstein B R 1983 Renormalization and radiative corrections at finite-temperature *Phys. Rev. D* **28** 340–8; Erratum **29** 3004 (1983)
[105] Weldon H A 1982 Covariant calculations at finite temperature: the relativistic plasma *Phys. Rev. D* **26** 1394–407
[106] Lee B W and Shrock R E 1977 Natural suppression of symmetry violation in gauge theories: muon- and electron-lepton-number non-conservation *Phys. Rev. D* **16** 1444
[107] Dolan L and Jackiw R 1974 Symmetry behavior at finite temperature *Phys. Rev. D* **9** 3320
[108] Lee T D and Nauenberg M 1964 Degenerate systems and mass singularities *Phys. Rev.* **133** B1549–62
[109] Kinoshita T 1962 Mass singularities of Feynman amplitudes *J. Math. Phys.* **3** 650
[110] Schwinger J 1961 Brownian motion of a quantum oscillator *J. Math. Phys.* **2** 407

www.ingramcontent.com/pod-product-compliance
Ingram Content Group UK Ltd.
Pitfield, Milton Keynes, MK11 3LW, UK
UKHW051857100926
8556IPUK00005B/665

9 780750 360524